Supersizing Science

Supersizing Science

On Building Large-Scale Research Projects in Biology

Niki Vermeulen

DISSERTATION.COM

Boca Raton

Supersizing Science
On Building Large-Scale Research Projects in Biology

Dissertation.com
Boca Raton, Florida
USA • 2010

ISBN-10: 1-59942-364-2
ISBN-13: 978-1-59942-364-7

The illustration on the cover, 'Spiral IV', is made by Kenneth Eward. I am grateful for his permission to use his work which stayed with me after I first saw it in the Science Visualization Challenge 2004, where it received an honourable mention:

> If you could climb the twisted ladder of a DNA molecule and look down, you might see something like the image above. Kenneth Eward, a science artist at BioGrafx Scientific & Medical Images in Ovid, Michigan, used x-ray crystallographic data from DNA molecules to paint a unique portrait of the double helix. The image omits the chemical bonds that crisscross the center of the molecule, so that the structural features of the helix can be seen more easily. (Grimm, 2004: 1905)

To me 'Spiral IV' does not only picture a vital part of modern biology, but it also represents the process of growing bigger. This idea, in turn, formed the basis for the cover design. I want to thank Jan van Beusekom for giving this Vermeulen book its cover.

Contents

Acknowledgements

Writing a book about scientific collaboration on your own may seem to be a mission impossible. Luckily, I did not have to do it alone. Although this book is the result of my personal research endeavour, I would like to acknowledge the support I received from many around me. First of all, I want to thank my supervisors Wiebe Bijker and Rein de Wilde. By showing me the delights of research Rein inspired me to engage in this study and Wiebe can be best described as my academic guide. You have given me full freedom to develop my own project, which has been a genuine learning experience. My fellowship at the University of York gave me a third person who has been vital. I want to express my sincere gratitude to Andrew Webster and all the other people at York who provided me with an academic home abroad. In addition, I am thankful for the support of my fellow PhD's and other colleagues from the Faculty of Arts and Social Sciences, the Netherlands Graduate Research School of Science, Technology and Modern Culture (WTMC), the European Association for the Study of Science and Technology (EASST), and the Society for Social Studies of Science (4S). Of course, I am especially grateful for the cooperation of the life scientists and policymakers who figure as key players in the subject of this study. Over the years I was fortunate to meet other scholars who investigate and explore scientific collaboration, and I am certainly looking forward to our future collaboration. Several persons I met during my research have meanwhile become good friends and are now part of the group of people who accompany me through life. My dear friends, let me assure you: your friendship is bigger than big! Finally, I would like to thank my close relatives who have been with me from the very beginning with love. Mam en Floor: samen staan we sterk!

Introduction

Nobel Prize winners James Watson and John Sulston both wrote memories of their contribution to molecular biology, respectively the reconstruction of the structure of Deoxyribose Nucleic Acid (DNA) in 1953 and the development and realisation of the Human Genome Project about fifty years later (Watson, 1968; Sulston & Ferry, 2002). Together, these books comprise an overview of major developments in the life sciences during the last century. At the same time a comparison of the two stories shows several striking differences. Watson tells about the scientific quest of a small group of scientists who pursue their research in a relatively small-scale academic environment. Informed by the work of others, Watson and Crick developed their model in the Cavendish Laboratory in Cambridge, the traditional English university town. Although Sulston's career began in the same academic setting – working on the genetic make-up of worms at the Laboratory of Molecular Biology – his story on the development of the Human Genome Project describes a completely different world. The deciphering of the human genome entails the planning and management of a large and dynamic project with a clear mission, involving huge amounts of money, expensive instruments and numerous scientists in laboratories all over the world. Moreover, the academic environment is substituted by an international and political setting, figuring academia, governments, funding bodies, business, media and the public.

What makes these two research endeavours in biology so different? How is it possible that Watson and Sulston are portraying such diverse pictures of a life devoted to science? The two stories indicate the emergence of a new way of doing research in biology. A central feature of the transformations in contemporary life sciences research is the movement towards more collaboration. While Watson and Crick worked together and only sometimes discussed their work with other scientists, the Human Genome Project connected forty-eight laboratories and involved more than five hundred scientists (Glasner, 2002). Moreover, the Genome project has been followed by various comparable large-scale research projects in biology (Eichinger, 2007; Glasner, 1996; Glasner, 2002; Editorial Nature, 2001). It is this transformation in the orchestration of contemporary life sciences research that I investigate in my thesis. I explore how in biology scientific and technological developments interact with organisational changes towards an increase in collaboration. I show how these collaborations are built and characterise scientific collaboration in biology.

My investigations into scientific collaboration in biology have to be understood in the context of studies on 'big science' and scientific collaboration in general. Interest in increasing dimensions of science began in the 1960s, when the term 'big science' was introduced (Price, 1963; Weinberg, 1967). Nowadays, science increasingly has become a collaborative effort, while also more and more studies of scientific collaboration have appeared (Hackett, 2005; Katz & Martin, 1997; Shrum et al., 2007). However, these old and new studies predominantly address disciplines that are known to have a tradition in collaboration, like astronomy, physics and space research. In contrast, I turn the spotlight on biology. Moreover, in the context of biology my thesis specifically focuses on the building of large-scale research projects in contemporary society. What kind of work has to be done to build collaboration? Adapting the notion of big science, I call this process of science becoming bigger the 'supersizing of science'.

Why is the supersizing of science and, more specifically, collaboration in biology worth investigating? Knowledge plays an important role in modern society and recent developments in the life sciences have a profound influence on diverse societal realms. As the desire to acquire knowledge, including knowledge about life, seems inherent to being human, research into life and its diversity has been fundamental in the development of modern science and the scientific profession (Pickstone, 2000). Nowadays, science has become a crucial component of society, with science being constitutive of many social processes, while it is almost impossible to think of society without taking the role of science into account (De Wilde, 2001; Jasanoff et al., 1995). More particularly, increasing knowledge about life at the molecular level has an enormous impact on agriculture, healthcare and industrial processes (Brown & Webster, 2004; Vermeulen, 2003; WRR, 2003). Think for example of genetically modified organisms, the development of genetic diagnosis and therapy, the use of biomaterials and bioenergy. Forensic research nowadays involves the examination of DNA as 'genetic fingerprints' (Butler, 2005).

Understanding science and the role of science in society, is the objective of a relatively new academic field that is known as the social studies of science, or

– including the role of technology – Science and Technology Studies (STS). Having its roots in the history, philosophy and sociology of science, this inter-disciplinary community of scholars studies the interaction between science, technology and society, finding that they are co-shaping in various ways (Jasan-off et al., 1995; Hackett et al., 2008; Sismondo, 2004). My thesis is written within this academic context, and I concentrate especially on the organisation of science. Social studies of science have shown that the organisation of re-search plays an important role in the creation of knowledge (Jasanoff et al., 1995; Knorr-Cetina, 1999; Latour, 1987; Vaughan, 1999). As a result, the building of collaboration in biology comprises a development that is worth reflecting on as it influences the construction of knowledge and is part of the supersizing of science.

How to study collaboration in biology?

One of the central insights from science studies holds that science has two faces: science in the making and ready-made science (Latour, 1987). Research – science in the making – is not an orderly process, which is planned at the beginning and followed by some logical actions that automatically lead to specific results. Often, it is only afterwards that science is presented as an orderly process: ready-made science. These insights into the ways in which science is made also poses a dilemma to the science studies scholar wanting to describe her own research process. Which road to take? Present ready-made science, or show science in the making? It will come as no surprise that to some extent my research process has been messy too. However, looking back it is possible to impose a certain sense of order on the processes involved. In this introduction I will recount both stories in a nutshell: first I will show some of the mess, after which I will explain how I managed to order it.

Beginning with the mess, let me tell about the origins of my research. This dissertation has its roots in Washington D.C., where I studied life sciences innovation policies for the Scientific Council at the Royal Netherlands Em-bassy. I found myself at the right place at the right time as it was the year 2003, marking the 50th anniversary of the publication of the structure of DNA by Watson and Crick together with the official completion of the Human Genome Project. To celebrate, lots of different activities related to the life sciences were organised. Most importantly the scientific symposium 'From Double Helix to Human Sequence – and beyond' at the National Institutes of Health, in which main actors in these developments, including Watson and Crick, told their stories about biology in past, present and future. This meeting was accompa-nied by a public symposium at the Natural History Museum and an exhibition on DNA. Moreover, there were policy oriented meetings on the Hill and the large annual BIO convention organised by the life sciences industry bringing together 16,234 participants from fifty-seven countries.[1]

In between attending these meetings and looking at science and innovation policies in the United States, I started wondering about this world of life sciences I was temporarily emerged in. Having a background in Science and Technology Studies, I recognised transformations that are described in recent literature on science and innovation. Like the emergence of what is called a new mode of knowledge production – the trans-disciplinary 'mode 2' research that takes place in the context of application – and the entanglement of academia, government and industry in the so-called 'triple-helix' (Gibbons et al., 1994; Etzkowitz & Leydesdorff, 2000; Nowotny, et al., 2001). However, what struck me most was the relative lack of attention to what all these changes meant for the actual research practice and the life of individual scientists. This made me decide that I wanted to look at contemporary transformations in the life sciences, focussing on the perspective of science and scientists.

Although my work can be best characterised as science studies, I did not perform ethnographic study of laboratory work, following scientists and describing their actions (Doing, 2007; Latour & Woolgar, 1986). Nor did I focus primarily on the analysis of science policy (Bucchi, 2004; Guston, 2000; Jasanoff, et al., 1995). Instead, I tried to find a place in between, where scientific practice and the organisation of science meet each other. In other words, I moved my study of science outside the laboratory into the space between laboratory practice and science policy. It is in this space that scientific collaboration is formed, entailing both scientific and organisational developments. Moreover, it is in this space outside the laboratory, where scientists are most clearly involved in and confronted with the contemporary societal attention to developments in biology, and the increasing entanglement of academia with government and industry. Finally, I pay explicit attention to the role of technology in science and the organisation of science, as scientific and technological developments go hand in hand in biology and other sciences that are characterised as 'techno-science' (Pickstone, 1993; Gibbons et al., 1994).

When doing laboratory studies, the sites of study are quite clear. But given that I decided to move outside of the laboratory, where, exactly, should my effort be situated? In fact, my empirical research took place at a variety of sites.[2]

---

[1] Number of attendees of BIO 2003 retrieved January 22, 2008 from www.bio.org/events/-2003/media/facts.asp

[2] Within anthropology this form of fieldwork is referred to as 'multi-sited ethnography' (Müller-Rockstroh, 2007).

First of all, I went to meetings, where scientists and other actors in the life sciences field meet. I visited scientific conferences, workshops, policy meetings, business events and network meetings and became involved in various networks of young scientists. Here I observed and – most importantly – talked to scientists and other actors actively involved in the world of life sciences, such as policymakers and businessman.[3] More specifically, I conducted in-depth interviews with a selection of those people, talking about their work and perspectives on developments in the life sciences, transformation in scientific practice and the role of scientists.[4] Secondly, I decided to concentrate on large-scale scientific collaborations in biology, because within these projects transformations in biology come together and materialise.[5] Moreover, these projects influence daily scientific practice and the professional life of scientists. I studied three different scientific collaborations in detail, analysing project and policy documents and talking to key scientists in the project, as well as to persons involved in the organisation or policies associated with the projects.

The projects I have chosen to study are all examples of contemporary scientific collaborations that investigate life, but in order to get an idea of both similarities and differences the three case studies represent diverse forms of research collaborations with different goals, varying from making an inventory, to transforming or applying information. First, the Census of Marine Life aims to study all animal life in the world's oceans and collect them in a virtual database. Second, the Silicon Cell initiative aims to build a model of a cell in a computer, synthesizing different data that have become known of various components of cells, including genetic information. Third, the VIRGO consortium is an academic-industrial collaboration that uses genomics techniques to develop new therapies against respiratory virus infections, such as influenza, and explicitly aims to apply knowledge that is acquired through research. I studied the various collaborations from their beginning until 2007, when I completed my empirical work.

---

[3] For an overview of attended meetings see Appendix A.

[4] For an overview of interviews see Appendix B. The quotes from Dutch interviews are translated into English by the author of this thesis.

[5] Scientific collaboration can have various forms (on definitions of collaboration, see Chapter 1 and the report of the 'Workshop Research Groups and Science Collaboration' by Braam & Verbree, 2008). I have chosen to concentrate on large-scale research projects, involving various national or international partners that present themselves as a research project and receive or aim to receive separate funding for this specific research project.

Presenting scientific collaboration in biology

In this thesis my research is arranged as follows. In line with the research process, I first make a general overview of transformations in biology research in preparation of analyzing scientific collaboration in detail. I employ the 'big science' concept to characterise transformations in contemporary life sciences research as big biology. By employing this concept, scientific and organizational developments in biology become part of a broader discourse on the expansion of science in modern society and the trend towards scientific collaboration. After placing developments in biology in a historic and cultural context, I characterize big biology as a specific networked form of big science. I show how scientific and technological developments in biology have interacted with social and political processes, resulting in increasing scientific collaboration. In contrast to more classic forms of big science like big physics, big biology has a networked character, which is underpinned by a socio-technical infrastructure with an important role for information and communication technologies. Moreover, I argue that big biology can be seen as a contemporary form of big science that corresponds with other trends in society and that indicates how big science in general has changed. In this context, I suggest that the analysis of big biology is exemplary of the broader trend towards scientific collaboration in contemporary science: the supersizing of science.

Supersizing focuses explicitly on the *process* of making science big. The increase of scientific collaboration in biology mirrors the building of scientific collaboration in contemporary society. My analysis of the three case studies not only shows how collaboration in biology has a networked structure, but it also illustrates how integration in biology takes place. By concentrating on the process of building big science I will explore how science is supersized and what kind of work is involved in the construction of scientific collaboration. I distinguish different styles of collaboration: projects have various rationales for collaborating as well as different orientations, resulting in diverse deliverables. I show how collaboration comprises different phases that require different types of work: the origin of collaboration, building connections and keeping it big. Within the process of building scientific collaboration, attention is paid in particular to what I call the 'projectification of science': the use of the project format to formalise scientific collaboration. I argue that as organizational structures, projects make it possible to connect scientific practice with government and industry, and that this directly affects scientific practice and the rhythm of science. In addition, I show how individual scientists play an important role in the shift towards the collectivization of science. In building scientific collaboration, scientists have to play new roles and in some respects become more similar to managers, politicians or businessmen.

Outline of the book

The book consists of three parts: a conceptual part, an empirical part and an evaluative part. The conceptual part, called "Big Biology" employs the term 'big science' to conceptualise transformations in biology. Starting with the debate on transformations in the organisation of research in the life sciences – asking whether biology is turning into big science or not – the first chapter *Big Science*. *Characterising transformations in science* explores the big science concept and compares it with other literature on scientific transformation and collaboration. The second chapter *Big Biology. Collaboration in the life sciences* compares developments in biology to traditional big physics. Some argue that biology is not big science, as technologies do not even come close to the size of instruments such as particle accelerators in high-energy physics, and consequently do not have the centralising force that constitutes big science complexes. However, I show how collaboration in biology differs from big physics on three fronts: it combines a different history of collaboration with the use of different kinds of technology and another social context. The chapter presents biology as a contemporary networked form of big science that emerged in interaction with the development of information and communication technologies.

"Life sciences live" – the empirical, second part of the book – takes the reader from the general conceptualisation of big biology to concrete examples of scientific collaboration. Subsequently, I present three different research projects that represent different styles of collaboration in biology. To explore these different forms of integration, I approach each case study from a different perspective, asking a different question. In the third chapter, *Seeing life in the oceans. New natural history*, I describe the Census of Marine Life as a contemporary example of natural history collaboration, taking a historical perspective to explore continuity and change. As the project can be compared to more traditional forms of scientific collaboration in natural history it enables me to study transformations in this style of collaboration. In contrast, the fourth chapter's Silicon Cell initiative is the result of contemporary developments in molecular biology, including the Human Genome Project, and shows how a new collaboration in biology is built. Next in *Growing a cell in silico. Constructing collaboration*, I analyse how the Silicon Cell project is deliberately staged as big science. By employing dramaturgical analysis – looking at collaboration as performance and the changing role of scientists – this chapter takes a sociological perspective on how collaboration is built. Finally, I investigate the building of a specific academic-industrial collaboration in the fifth chapter, *Developing a new vaccine. The innovation epidemic*. Since the VIRGO consortium is explicitly put forward as an 'innovative cluster' in which academia and industry work together, it becomes possible for me to investigate the actual formation of academic-industrial collaboration. Building on theories of innovation, I will explore how academia, government and industry become intertwined. As a result, the case studies represent three different processes that are fundamental to the construction of contemporary scientific collaboration in biology: the transformation of traditional collaboration, the building of new collaboration and the construction of academic-industrial collaboration.

The empirical part of this thesis provides the foundation for the final, evaluative part, entitled "Supersizing science". In this conclusion I pay attention to the structure and dynamics of scientific collaboration. Based on the empirical findings, the sixth chapter, *Unpacking collaboration in biology. Supersizing science*, analyses different forms of networks in biology. After looking into the structure, size and style of collaborations, the process of supersizing science is put centre-stage. I describe the work that has to be done to build large-scale scientific projects in biology and relate this to the growing importance of scientific collaboration in general. In addition, I propose to put the 'projectification of science' on the agenda of science studies. Finally, I turn the attention to networks of young scientists in my epilogue, *The future of science. The next generation*.

PART 1

Big Biology

CHAPTER 1

Big Science
Characterising transformations in science

"Life sciences is developing very rapidly into 'big science', with huge effort invested in it worldwide". This sentence can be read on the website of the Swammerdam Institute for Life Sciences of the University of Amsterdam.[6] Under the banner *Life Sciences, the science of this century* the institute presents itself as part of a global life sciences community that is nowadays changing fundamentally and becoming more prominent. A focus on the micro-level of molecules is providing new perspectives on life and increases the importance of the biosciences. Biology is very much alive, not only within the community that investigates life, but also in business, politics and media. According to geneticist Maynard Olson "The change is so fundamental, it is hard for even scientists to grasp" (Roberts, 2001: 1182). The term 'big science' is used to capture the enormous expansion of biology. More specifically, the Human Genome Project is considered the first big science project in biology (Kevles & Hood, 1992; Lenoir & Hays, 2000; Venter, 2007).

By employing the concept of big science in the context of biology, transformations in the organisation of research are framed in terms of growth and the movement towards collaboration. To illustrate, general features of big science are increasing numbers (money, manpower) as well as an increase in multi-disciplinary collaboration, the use of large, expensive instruments, the industrialisation of research, the tightening of relations between science, government and industry and internationalisation (Galison & Hevly, 1992; Sklair, 1973). In addition, by employing the term big biology, transformations in the life sciences are compared to similar developments in other scientific fields that are conceptualised as big science, most notably large-scale research projects in physics. The term big biology is especially employed by actors in the life

---

[6] Retrieved in 2004 from http://www.science.uva.nl/sils

sciences field. Big biology is used as a rhetorical weapon in the debate between scientists and policymakers in favour of the Human Genome Project and those opposing the project.

This *Big Biology* section employs big science as an analytical concept to study transformations in biology. I will show how the use of the big science concept helps to explore scientific and organizational developments in biology and places these transformations in a historical and cultural perspective. In order to characterize big biology, I will first examine the big science concept and its value as an analytical concept. Starting with the analysis of the debate on big biology, the first chapter will look into the origin and the history of the concept and distinguish different meanings of the term. These different meanings reflect the multi-faceted character of increasing dimensions in science and the various perspectives from which it can be studied. In this respect, the second chapter will analyse collaboration in biology as a form of big science.

Debating big biology

The Human Genome Project was accompanied by a heated debate about big biology. Some argue that the project can be seen as the start of big biology, while others firmly deny that biology is turning into big science (Check & Castellani, 2004; Davis, 1990; Rechsteiner, 1990; interview Remacle, 2006; Roberts, 2001: Venter, 2007).[7] In defence of big biology, the Genome Project is compared with the Manhattan Project that developed the atomic bomb during World War II, which is one of the biggest science projects in the history of physics (Lenoir & Hays, 2000). In contrast one of the headers in a discussion about the Genome Project clearly states: "The HGP Isn't 'Big Science'" (Hood, 1990: 13). Arguments in the discussion focus on the amount of money spent and the project management structure. Those defending big biology argue that the almost 3 billion dollar budget of the Human Genome Project is enough to make it a big science project, while for others biology does not qualify as big science because investments are small compared to physics or research into space. Furthermore, a central mode of organisation – as in big physics research – is seen as distinctive of big science endeavours, which turns the relatively fragmented, international structure of the HGP into an argument against the label 'big'. Since the 1990s both parties have stuck to their arguments and it is

---

[7] Personal communication with Prof. dr. André Goffeau, Institut des Sciences de la Vie, Université catholique de Louvain, Louvain-la-Neuve, Belgium: January 11, 2007.

safe to argue that the debate on transformations in the organisation of the biosciences has not reached any definitive conclusions yet.

However, careful analysis of the debate will reveal that the term big science is used as a symbol for a new, modern and more industrial way of organizing research. The big science concept is not primarily used to analyse the increasingly large dimensions in research, but it presents the Human Genome Project as a totally new way of doing biology. Next to new developments in technology, a specific mode of management has come into play, whereby scale, efficiency and work division are central. Moreover, authorship becomes collective. This new big science style of research sharply contrasts with more traditional images of science, symbolized by big science's antonym 'little science'. Small-scale science portrays research as an individual creative endeavour, resembling craftsmanship instead of an industrial way of working. This more romantic view of science also entails the image of the genius or distracted professor working alone in his study or laboratory (Doorman, 2004). In the case of biology, little science often has the character of investigator-initiated research in small laboratories (Roberts, 2001).

The debate on big biology, then, is about different ways to organise life sciences research, while the big science concept is mainly employed as a rhetorical weapon. Big biology has become a central phrase in the battle between scientists in favour of the Genome Project and those opposing it. At stake is not only the actual character of biology research but also the shaping of the future of science. Do we want the future of biology research to be little or big? Proponents of the Human Genome Project present big science as the new and more effective way to perform research: "scientific leaders agree that collaborative projects can produce results that would be impossible for specialized individuals working alone to achieve" (Check & Castellani, 2004: 546). In contrast, opponents of the Genome Project state that big biology undermines the very character of science, because it industrialises, bureaucratises and politicises research and dilutes creativity. To illustrate, genome sequencing is portrayed as 'massive, goal-driven and mind-numbingly dull' (Roberts, 2001). Molecular biologist Sydney Brenner even joked that sequencing is so boring it should be done by prisoners: "the more heinous the crime, the bigger the chromosome they would have to decipher" (Roberts, 2001: 1183).

As a result, it seems, the big science concept has provided discussants with a strong rhetorical sword for strengthening their case and draw a clear demarcation between traditional and new forms of organising life sciences research. But big science turned out to be a double-edged sword: it has both positive and negative connotations and could therefore be used by proponents as well as opponents of the Genome Project. Within the debate, big science is also put forward as an empirical concept, however. Whether or not the Human Genome Project should be considered big science became a major concern in the debate. Although this question could have given rise to a detailed analysis of transformations that take place in biology, the question was put forward in the context of the debate and was only answered strategically. Opponents of the Genome Project considered it big science while making use of negative normative connotations of the term and stressing discontinuity with normal research practice in biology. In contrast, the project's proponents emphasized the positive connotations of bigness or tried to win the argument from opponents by empirically denying the big science character of the Genome Project.

In sum, this analysis of the debate on big biology explicates the different ways in which the term big science is used. While life scientists and policymakers employ the different faces of big science, they do not reflect on the origin and the meaning of the concept and therefore do not explore the meaning of big biology in detail. For instance, it was not taken into account that bigness is of course relative (NRC Committee on Solar-Terrestrial Research, 1994), or that bigness can be found in various dimensions: geographic, economic, multidisciplinary and multinational (Galison, 1992). These different meanings that are inscribed into the big science concept only become evident when taking a closer look at the concept and its history. What does the concept cover? This is explored in the next section by investigating the big science concept as a starting point for discussing transformations in biology in a more nuanced way.

The meanings of big science

Opening the black-box of big science constitutes a starting point for exploring the way in which the concept may help us to grasp contemporary transformations in biology. By performing a conceptual analysis of big science, I will show how this concept is more then a rhetorical device in the fight between opponents and proponents of the Human Genome Project. Introducing the concept of big biology adds a specific view on transformations in science that can ultimately lead to a better understanding of contemporary developments in biology. To enfold the various meanings of big science, I will go back to the roots of the concept and discuss its development.

The origin of big science

The history of the term big science starts in 1961, when it was first introduced by Alvin Weinberg in his article on the *Impact of large-scale science on the United*

States (1961). Later, in his book *Reflections on big science* (1967) he defines the term as follows:

> Science has become big in two different senses. On the one hand, many of the activities of modern science – nuclear physics, or elementary particle physics, or space research- require extremely elaborate equipment and staffs of large teams of professionals; on the other hand, the scientific enterprise, both Little Science and Big Science, has grown explosively and has become very much more complicated. (Weinberg, 1967: 39)

Weinberg's attention to growth in science can be traced back to his participation in Americans large nuclear energy projects and his interest in science policy. In 1946 he became director of the Physics Division of Oak Ridge National Laboratory (ORNL) and in the period from 1955 till 1973 he headed the National Laboratory, the largest physics plant in the world and the largest factory of any type anywhere (Galison & Jones, 1999). However, Weinberg was more than a laboratory director: "In approximate chronological order, he was a physicist, a pioneer in nuclear energy, a reactor builder, a laboratory director, a lecturer, a writer, a thinker, a policymaker" (Zucker, 1995: 5). It is in this last role – as member of the President's Science Advisory Committee and director of energy R&D in the White House – that he came up with the term big science and composed his study.

In the preface of the book that comprises a collection of talks he gave during his years as laboratory director, Weinberg explains what inspired his reflections. First, he positions himself as director of a Big Science institute, a special feature of modern science in contrast to the traditional university. As director he feels obliged to justify the existence of the national laboratories and "the large sums of public money that pour into modern science" (Weinberg, 1967: v) and he also wants secure support as "the first job of a laboratory director is to assure continuing and ample support of the institution he directs" (idem). However, in the second half of the book Weinberg changes hats and becomes science advisor. Stating that the competitive and formal atmosphere in science committees in his opinion "can no more produce wisdom than design a camel" (Weinberg, 1967: vi), he gives his own view on the most troubling questions of modern science: the allocation of resources among competitive scientific fields and between science and other public enterprises, and "whether the new style of Big Science is blunting science as an instrument for uncovering new knowledge" (idem). With his book Weinberg aspires to provide a common language and framework for discussing issues related to the growth of science.

When addressing the origin of big science, however, most people will refer to the classic book of Derek de Solla Price: *Little Science, Big Science* (1963). This author elaborates on the concept of Weinberg, writing enthusiastically: "The large-scale character of modern science, new and shining and all powerfull, is so apparent that the happy term "Big Science" has been coined to describe it" (p. 2). Price started his career as a physicist, but as he explains it was a pile of journals that triggered his interest in the history of science and his engagement with big science (Price, 1983). Teaching at Raffles College – now the University of Singapore – he received a complete set of the *Philosophical Transactions of the Royal Society of London* from 1662 to the 1930s:

> I took the beautiful calf-bound volumes into protective custody and set them in ten-year piles on the bedside bookshelves. For a year then I read them cover to cover, thereby getting my initial education as a historian of science. As a side product, noting that the piles made a fine exponential curve against the wall, I counted all the other sets of journals I could find and discovered that exponential growth, at an amazingly fast rate, was apparently universal and re-markably longlived. (idem: 18)

While within the community of historians of science Price stood alone with his fascination for growth, his ideas were enthusiastically received by physicists during the Pegram Lectures he gave at Brookhaven. He rewrote those lectures into his book on big science which became an immediate success and con-nected very well with two new emerging fields in the 1960s: the sociology of science and information science.

Although Weinberg and Price belong to the same generation of physicists and share their fascination for big science, they engage with the big science concept in different ways. While Weinberg explicitly uses the concept to iden-tify and evaluate transformations in modern science, Price's elaboration of the concept is more descriptive and focuses especially on the quantitative growth of science. However, in both accounts the emergence of big science is received with mixed feelings. On the one hand, the books can be read as celebrative descriptions of the growth of science as part of the development of science and society. On the other hand, the authors also critically reflect on the implications of big science. Particularly Weinberg perceives the growth of science as a problem, asking for example whether it is a good way to perform research or if investments in physics could not be better spend in other parts of society. In addition, Price contemplates the limits of growth. He observes the problems the growth of information poses and he takes the changes it requires, such as specialisation and teamwork, into account. He also suggests that scientists will lose their 'mavericity' (the ability to think outside the box and make unconven-tional associations) when working in large-scale research projects. This ambiva-lent stance towards big science, which is still very much visible in the two opposing views on big science in the debate on big biology, can be understood in the context of Modernity, in which the concept emerged.

Big science as a modern concept

The first books on big science are part of a long list of books with the term 'big' in the title that address growth as a distinctive phenomenon of modern society: *Big Business* (Fay, 1912, Hendrick, 1919, Drucker, 1947), *Big Government* (Pusey, 1945), *Big Democracy* (Appleby, 1945), *Big School* (Barker & Gump, 1964), *Big Cities* (Rogers, 1971), *Big Foundations* (Nielsen, 1972) and *Big Machine* (Jungk, 1986). These books all focus on overwhelming growth in modern society and its supposed inevitability. They focus on the implications of growth for the organisation of society and for individual man, often with a critical normative connotation. Growth is not only seen as an exponent of modern industrial society, but also as a source of its problems. Drucker (1947) even talks about the "Curse of Bigness" (p. 211). Within this tradition of 'big books' science is just the next sector of society that is described in terms of bigness, paying attention to scale and its positive and negative implications.

The term "big" is very much connected to the United States, the country of the Big Mac hamburger. So it is not surprising that it is also the place of birth of the big science concept and the home country of most of the other big books. When looking at the origins of the term Capshew and Rader observe: "It is no accident that the existence of Big Science was first discerned in the United States, where growth is a way of life and bigger is often viewed as better" (Capshew & Rader, 1992: 3). More specific, Galison (1992) places the roots of the development towards big science in the context of the Great Depression of the 1930s. In the United States it caused a counter reaction that admired bigness and gave rise to enormous building projects, such as Golden Gate Bridge, Hoover Dam and Empire State Building: "Without the cultural fascination of Americans in general for the large, the goal of building ever larger scientific facilities might have remained peripheral to other concerns" (p. 3). However, these writers on big science do not explicitly connect the origin of the concept to its cultural background of modernity.

The big books are written in the context of the broader cultural development of modernisation. In Modernity, growth is accompanied by the ordering and (re-)structuring of society (Berman, 1983; Kumar, 1995; Latour, 1993). Rationalisation and industrialisation processes give rise to so-called Taylorism and Fordism – efficiency and scaling-up – and are accompanied by bureaucratisation. Apart from the possibility of structuring and organising society efficiently, people strongly believed in progress. However, the modern condition

23

also involves awareness of and reflection on the changes that take place. In these reflections modern life is fully embraced and at the same time harshly criticised: "To be modern is to find ourselves in an environment that promises us adventure, power, joy, growth, transformation of ourselves and the world- and, at the same time, that threatens to destroy everything we have, everything we know, everything we are" (Berman, 1983: 15). The modern order is even pictured as an iron cage (Weber, 1958). This critical stance towards the development of bigness is clearly visible in Weinberg's *Reflections on Big Science* (1967) and ultimately illustrated through the book *Small is beautiful* (1973) written by economist E.F. Schumacher. As this brief overview reveals, the context of modernisation is fundamental for the positive and negative normative connotations that are still connected to the big science concept today.[8]

Finally, the emergence of the term big science has to be placed in the context of science policy after World War II. Although the Manhattan project is frequently named as reference point for large-scale science, the big science term was coined *after* the war, in American society of the 1960s. In this postwar period important transformations took place in the relationship between science and society. During wartime, governmental investments in science increased enormously, but they also strengthened the political grip on the direction of scientific research. The ending of the war was therefore seized as an opportunity to renegotiate the relation between government and science (Galison, 1992; Guston, 2000; Jasanoff, 2005; Rip, 1998). The social contract for science based on the famous report of Vannevar Bush (1945/1980) can be read as a way to (re-)establish the divide between government and science after the War. He wanted to safeguard the continuity of government investment in science while also creating a protected space for science by putting it forward as an investment that always pays off, using successful scientific applications of the war as a symbol for the societal usefulness of basic science. Writings on big science reflect this view on science; fundamental science and the application of science go hand in hand and do not exclude each other. Moreover, the clear divide between policy, science and society coincides with the first generation of

---

[8] Currently, similar discussions surround the concept 'supersizing' that seems to be the new 'big'. The term 'supersize' also originated in America. It was first used in 1917 by a manager of the Firestone Tire and Rubber company, announcing the production of a new type of tire: the Firestone supersize cord tire (The Oxford English Dictionary Online. Retrieved June 11, 2008 from http://dictionary.oed.com). Recently the term became popular in the fastfood industry, which supersizes meals and the critical documentary about this process made by journalist Morgan Spurlock called 'Super size me' (2004) (Retrieved June 11, 2008 from http://www.supersizeme.com/). However, the term 'supersize' is also used in other contexts to indicate the process of becoming bigger, having a similar normative connotation as the term 'big'.

big science writers in which scientific practice – the actual research process – is not discussed.

The development of the big science concept

What happened with the 'happy term' after the 1960s? Common to early definitions of the concept is their focus on the description of increasing scales in science, measured in terms of money, scientists and publications. For instance sociologist of science Sklair defines big science short and simple "Big Science: Money and Manpower" (1973: 15). However, in later definitions different dimensions of growth are added. For instance, Lambright includes a timescale: "Big Science"-large-scale research and development (R&D) programs costing hundreds of millions, even billions, and lasting a decade or more" (1998: 260). In turn, Galison extends the dimensions of growth further again:

> [T]he big in Big Science connotes expansion on many axes: geographic (in the occupation of science cities or regions), economic (in the sponsorship of major research endeavors now costing in the order of billion dollars), multidiscipli-nary (in the necessary coordination of teams from previously distinct fields), multinational (in the coordination of groups with very different research styles and traditions). (1992: 2)

As a result, definitions of big science now indicate expansion in many directions: increasing numbers of scientists and publications, increase of investments in science, growth of scientific institutions, development of large instruments, increasing multi-disciplinary and multinational scientific collaboration, geographical expansion of science and increasing duration of scientific programmes.

Especially the period at the end of the 1980s and the beginning of the 1990s has been important in the development of the concept and its broadening meaning. Various publications on big science appeared and the highlight of this period is a conference at Stanford University in 1988, which resulted in the edited volume *Big Science; The Growth of Large-Scale Research* (Galison & Hevly, 1992). The preface of the book reports an interdisciplinary workshop with a diverse mix of scholars that employ the big science concept to look into a variety of subjects.

> Where past studies on big science typically counted dollars and personnel, and tabulated the funding sources that nourished large-scale research, we can now see more of the causes and consequences of the growth of science. (Hevly, 1992: 357)

This volume presents case studies of various large-scale research endeavours, paying attention to the use of complex and expensive technologies, multi-disciplinarity, the role of government, the funding and management of science, bureaucracy, the position of scientists, creativity, the intermingling of different actors (government, academia, industry), the organisation of private R&D and the emergence of science regions. In addition, different scientific realms are added to the initial focus on physics research.

Mapping big science

In line with Weinberg, physics research around World War II is generally seen as the exemplary situation for the big science concept (Capshew & Rader, 1992; Galison & Hevly, 1992; Nauta, 1984; Weinberg, 1967). However, it is possible to date back big science practices earlier than World War II. Historians of science frequently name astronomy with its large telescopes and observatories as the earliest form of big science (Capshew & Rader, 1992; Price, 1963; Smith, 1992). Price presented the origins of big science in astronomy:

> [T]he great observatories of Ulugh Beg in Samarkand in the fifteenth century, of Tycho Brahe on his island of Hven in the sixteenth century, and of Jai Singh in India in the seventeenth century each of which absorbed sensibly large fractions of the available resources of their nations. (Price, 1963: 4)

Building on this, historian of astronomy Robert Smith observes that the use of larger telescopes in the nineteenth century and the consequent demands of construction, maintenance, administration, and finance brought a social process of rationalisation and an industrial organisation model (Smith, 1992). Leaders in the research community instituted an increasingly differentiated and hierarchical division of labour within the astronomical workforce (Capshew & Rader, 1992).

In addition, historical observations discard the dominance of large instruments in big science, as large-scale projects with a complex infrastructure are also described as big science. For instance, the grand alliance between science and exploration in the 17th century is a source of big science constellations:

> In 1761 and again in 1769 scientists from England, France and several other countries put together expeditions to observe the transit of Venus across the face of the sun in order to refine measures of solar parallax. These endeavours required major investments in equipment and logistical support and contributed to the evolution of an international community of scientists. (Capshew & Rader, 1992: 21)

Infrastructural developments to transport people and information were crucial in this form of big science. Natural historians were able to collect more information with the help of research assistants in the field, and both Linnaeus and Darwin were embedded in extensive correspondence networks. In addition, large-scale topographical projects, such as the British 'Magnetic Crusade', created a worldwide network of geomagnetic observatories in the 1830s that were part of this category of big science. In contrast to vertical integration (physics mode of centralised big science), big science can thus entail horizontal integration as well – consisting of scattered sites of inquiry connected through a network of transport and communication infrastructure.

The search for the historic roots of big science practices also sheds new light on the origins of big physics: "If we seek the origins of Big Science in physics, we must look to the period between the two world wars" (Seidel, 1992: 21). It is argued that big physics emerged in California where universities such as Stanford, Caltech and UC Berkeley became involved in the problems of power production and distribution (Galison, 1992; Galison & Jones, 1998). In the 1930s Lawrence's Radiation Laboratory in Berkeley became the centre of energy research, housing the first cyclotron, according to the Rockefeller foundation: "a mighty symbol, a token of man's hunger for knowledge" (Galison 1992: 3). This laboratory grew into a national and international centre of nuclear science, where physicians, biologists, chemists and engineers worked together on the cyclotron to manufacture substances and propagate beams for use in experiments and therapy. Radar, counter-radar and atom projects would spread the Californian big science mode of working to the rest of the United States.

This context of energy research also shows industrial collaboration in big science constellations (Galison 1992; Galison & Jones, 1998). Resources for energy research not only came from the federal state, the state of California and a philanthropic organisation like the Rockefeller Foundation but also from industry. This was also the case with the emergence of microwave technology at Stanford. Apart from particular merits of public-private cooperation, studies of industrial collaboration within big science also reveal several concerns, such as secrecy and the fixation on patenting. The character of research starts to change: "instead of exploring new phenomena, the physicists found themselves increasingly spending their time searching for ways to pursue patentable ideas" (Galison, 1992: 4). Although in World War II extensive government investments in research offered an alternative to private financing and the patenting business, this brought restrictions on disclosure for reasons of national security and limits placed by technological needs of the armed forces.

Post-war science continued in the spirit of World War II experiences (Galison, 1997). In the United States the weapon projects became symbols for the future of physics research and other scientific fields:

> "[T]he war trained academic physicists to think about their research on a new scale, invoking a new organizational model (...)The Manhattan Project was far more than an indicator of the usefulness of physics; it was a prescription for the orchestration of research. (Galison, 1997: 305-309)

Similarly, big physics entered the stage in Europe with the establishment of the European Organization for Nuclear Research (CERN) in Geneva in 1953 (Pestre & Krige, 1992). In addition, space science expanded in the context of World War II and the Cold War that followed (Graham, 1992; NRC Committee on Solar-Terrestrial Research, 1994). Radar and missile projects built on early large-scale explorations of space. In the context of the 'space race' the United States initiated the Apollo programmes that became institutionalised in the National Aeronautics and Space Administration (NASA) in 1958. These endeavours were mirrored by similar ones in the Soviet Union as well as in Europe. This new way of organising research was envisioned to extend to other scientific fields, like chemistry, biology and medicine (Price, 1963; Galison, 1997).

Finally, recent research presents detailed case studies of different forms of big science in fields as diverse as astronomy, ecology, physics and space research and enriched the empirical understanding of big science (Bocking, 1997; Crease, 1999; Galison, 1997; Hoddeson et al., 1993; Kwa, 1987; Lambright, 1998; Schloegel and Rader, 2005; Westwick, 2003). In sum, the emergence of large-scale research complexes became perceived as a general trend and although scientific fields have their own characteristics, they can be presented as different forms of big science.

New perspectives on (big) science

These studies into the history and development of big science complexes do not only trace large-scale scientific practices from their emergence in astronomy to contemporary research endeavours, they also indicate developments in the study of these practices. When Price called himself a historian of science, he referred to a way of studying the history of science that differs considerably from ways in which science and its history is studied nowadays. When Price was introduced to the history of science it was a relatively young discipline that primarily studied the history of scientific instruments and the scientific revolution as a turning point in history that fundamentally changed people's outlook on the world (Mayer, 2000). In contrast, big science studies from the end of the twentieth century and the beginning of the new millennium present detailed studies of scientific practices. This shift has taken place in interaction with transformations in the study of science. The history of science as known by Price and the 'science of science' that emerged after World War II ultimately gave rise to the social studies of science and technology. Now science is perceived as a social system, changing the ways in which scientific practice is studied, as well as the phenomenon of big science.

During the period in which the big science concept emerged, the character of the study of science started to change (Edge, 1995; Jasanoff et al., 1995; Cutcliffe & Mitcham, 2001). A sociological perspective on science was added to the traditional history and philosophy of science, perceiving science as a social system that has to be studied accordingly. In the context of debates on the relationship between science and society and the search for a rational basis for science policy in the 1960s, a practical need for a 'science of science' appeared. This new perspective on science also has important roots in the critical outlook on science that can already be found in Weinberg's book on big science. Not only the relation between science and society became subject to debate, but also the social responsibility of scientists became a topic of discussion. In addition, the connection between technology and power was scrutinized against the backgrounds of the disastrous effects of atomic bombs, debates on nuclear energy and the horrors of the Vietnam War. These developments ultimately led to a new perception of science and technology. The linear, truth-finding image of scientific progress is now challenged by a constructivist perspective on science, claiming that science and technology are socially constructed and also incorporate relations of power.

After the politically inspired start these social studies of science and technology, grew into an academic discipline called Science and Technology Studies (STS) which primarily studies the place of science and technology in society, without direct political aspirations (Bijker, 2001). Although strong voices currently argue for a (re)connection to practice and a strengthening of the political dimension of Science and Technology Studies (Hackett et al., 2007), this theoretical orientation – which Bijker called a detour into academia – gave the field a stronger academic profile. The empirical employment of the big science concept at the end of the 20th century can be seen in the light of these developments. The various recent studies into big science employ the concept to describe the expansion of science in an academic context, investigating scientific practice and the relation between science and society empirically, combining recent insights from the history and sociology of science. Hence these diverse studies of big science are not only influenced by the transformation of science studies; they have also been important for the development of this new academic field that tries to understand science.

An overview of the different meanings of big science is exemplary for this academic turn in the history of the big science concept. The concept is used as a tool for looking into large-scientific endeavours and this also brings reflection on the concept itself. The article "Big Science: Price to the Present" (Capshew & Rader, 1992) gives an extensive summary of what big science entails. The authors carefully deconstruct the term's different uses, which show the influence of the development of science studies on the concept. First, the article focuses on the early meanings of the big science concept. *Big Science as a pathology* is associated with Weinberg's critical outlook on the state of contemporary scientific institutions. It views big science as a phenomenon that is ruining science. It covers the terminal diseases of mature science: journalitis, moneyitis and administratitis. Conversely, *Big Science as a scientific phenomenon* goes back to Price's first attempt to chart the overall historical growth of science by means of a variety of statistical indicators. Next the technological and political dimensions of big science are explicated. *Big Science as an instrument* refers to monumental technologies as hallmark of big science. In addition big science can also be a political instrument. NASA, for example, can be seen as a political agency taking political decisions: a state within the state. *Big Science as politics* focuses on sciences relation to government, democratic control over science, accountability and the intellectual autonomy of science.

Furthermore, Capshew and Rader present big science as a concept relating to scientific practice and the organisation of science. *Big Science as industrial production* focuses on the organisational consequences of the expansion of scientific practice. These organisational consequences are often conceptualized as the industrialisation of research, referring to hierarchical structures, teamwork, the division of labour and the commoditization of knowledge. In short, big science becomes synonymous to 'think factories'. *Big Science as an institution* concerns the institutionalisation of big science and the entanglement of scientific, technical, social, institutional, economic and political dimensions in different local contexts. *Big Science as an ethical problem* deals primarily with the internal ethical standards of science, the (decreasing) autonomy of scientists, the (increasing) dependency on patrons, and the choice of research priorities. *Big Science as a form of life* is a more holistic approach that connects the different aspects of big science into what Pickering calls a form of life: "the emergence of a unitary big-science constellation of scientific-technical-political-institutional practice" (Capshew & Rader, 1992, 17: citing Pickering, 1989). Finally, *Big Science as culture* refers to the great cultural significance of science in western society and the intimate relationship between science and society, a basic presumption of current studies of science and technology. [9]

With this overview Capshew and Rader not only give insight into the various perspectives on big science, they also intervene in the meaning of the term. Reflecting on the value of the big science concept, the authors present growth as the most important aspect of science: "the growth of science is perhaps its most notable historical characteristic, whether considered in terms of scope, scale, complexity, or impact" (Capshew & Rader, 1992: 3). However, they notice that nowadays big science is often viewed as somehow distinctive from the historical processes that produced it. This disconnection, they argue, results into the conventional 'mnemonic of money, manpower, machines, media, and the military' and an exclusive concern with the attributes of bigness, drawing attention away from what they see as "the more significant and interesting question of how science becomes larger" (p. 4). Consequently, they propose to move away from the more qualitative description of science *being* big to looking at science that *becomes* big(ger) in order to get "a better understanding of how science is made big and how 'big science' is made" (Capshew & Rader, 1992: 25). It is this process that I have called the 'supersizing of science' which I will study in the context of biology.

New perspectives on science and collaboration

So far this chapter has shown how big science and its antonym little science form a dichotomy that is used for diagnosing and discussing transformations in science, related to the increase of scale. I have shown how the big science concept should be seen as a historical concept that has developed over time, thereby taking on different meanings: big science has an empirical as well as an evaluative side, and both have a double edge. Coined in the 1960s, big science is originally a modern concept that evaluates the modernisation of science positively as well as negatively. When looking at the empirical side of big science, a division can be made between a quantitative and a qualitative perspective. Price used the concept to measure increasing scales in science, while in the context of historical and social studies of science the concept is also employed to study transformations in science in a qualitative way. In addition the concept is put forward to investigate the supersizing of science. Nevertheless, the use of the

---

[9] They even point out that big science has its own song, one by the avant-garde performance artist Laurie Anderson (1982).

concept to study increasing dimensions in contemporary science, including modern biology, is not self-evident because big science is often replaced by new dichotomies for studying transformations in science and new perspectives on scientific collaboration.

New dichotomies for studying transformations in science

In recent years, new dichotomies have taken the place of little and big science. Most notably, transformations in knowledge practices are characterised as transitions from 'industrial' to 'post-industrial' society; from 'mode 1' to 'mode 2' science; from separate societal domains to the 'triple helix'; and from 'normal' to 'post-normal' science (Bell, 1967; Drucker, 1969; Gibbons et al., 1994; Funtowicz and Ravetz, 1993; Funtowicz and Ravetz, 1994; Leydesdorff & Etzkowitz, 2001; Nowotny et al., 2001). These concepts have the same dual function as big science: they are geared towards diagnosing and discussing change. They diagnose the current situation and its characteristics in contrast to the older situation and thereby also clearly intervene in our thinking about changes in science and society. The different framings of the contemporary situation all stress different aspects of science, appreciate specific elements while declining others and therefore imply specific directions for future action.

The transformation from an *industrial* to a *post-industrial society*, as suggested by Bell (1967) and Drucker (1969), introduces a new type of society: a knowledge society in which science and technology follow industrialisation as the foundation of Western society. While the diagnosis of both authors converges, the implication for the position of science in society is envisioned in different ways (De Wilde, 2001). Bell promises a privileged place for theoretical knowledge and knowledge institutions, while Drucker predicts the industrialisation and commoditization of knowledge. This last stance is reflected in contemporary dichotomies. The transformation from *mode 1* to *mode 2 science* proposes a new form of knowledge production that displaces disciplinary and fundamental knowledge practices with a more reflexive, transdisciplinary and heterogeneous 'knowledge production' that takes place in the context of application (Gibbons et al., 1994). Accordingly, the authors identify and more or less advocate the reform of established institutions, disciplines, practices and policies. This message is also powerfully implied by the *triple-helix theory*, which signals an organisational change from separated domains in society to the entanglement of the domains of science, government and industry (Leydesdorff & Etzkowitz, 2001). Especially from a policy perspective this transformation towards a triple-helix can be read as a recipe for innovation.

Today's dichotomies focus also on the embedding of science in society. Some propose the addition of the public as a fourth helix (Etzkowitz & Leydesdorff, 2003). Moreover, the relation between science and society is central in a second book on mode 2 science (Nowotny et al., 2001). Against the background of a transformation to a mode 2 society – characterised by increasing complexity, uncertainty and reflexivity – the authors make a strong case for a more thoroughly embedded science through the contextualisation of knowledge in society, leading to what they call 'socially robust knowledge'. Similarly, Funtowicz and Ravetz (1993; 1994) identify a transformation from *normal* to *post-normal science* in the context of growing uncertainty. They state that when uncertainty becomes fundamental and risks are high, knowledge practices take a different form from what Kuhn named 'normal science'.[10] Post-normal science incorporates not only scientific methods and values, but is also shaped by values and deliberations from outside the scientific community, e.g. government and the public. Consequently, Funtowicz and Ravetz argue for a more pluralistic strategy of inquiry in which quality assurance takes the form of 'extended peer review'.

When surveying these new dichotomies that describe and evaluate transformations in science and society today, it becomes apparent that they do not discard the old dichotomy between little and big science. Actually, they often present characteristics that are already manifest in the big science concept. A comparison shows remarkable resemblance, in fact, notably pertaining to the increasing importance of knowledge in our society, the industrialisation of knowledge, the emphasis on application of knowledge and the relationship between science and society.[11] Although in the context of contemporary society the outlook on these issues has changed, it can be argued that the dichotomy between little and big science already entails more contemporary dichotomies to some extent. In addition, transformations related to increasing scales are still very relevant in science today. For instance, it is worth noting that the recent dichotomies are published by multiple authors while the first books on big science still involved single authorship. And this is only one example of how big science becomes apparent in more and more scientific fields. However, recent dichotomies do not focus explicitly on increasing dimensions in science and therefore they do not replace the dichotomy of little and big science.

---

[10] See also "Post-normale wetenschap in actie. De rol van veldonderzoek in het Britse debat over genetisch gemodificeerde gewasteelt" (an essay written by Wilde, R. de & M. Reithler for the NWO programme on Ethics, Research and Governance).

[11] For an overview of commonalities and differences between these contemporary perspectives, see also Hessels & Van Lente (2008).

Next to new perspectives on transformations in science, new ways for studying scientific collaboration have appeared. In recent literature on scientific collaboration various definitions and approaches to study the phenomenon can be recognised (Chompalov et al. 2002; Chompalov & Shrum, 1999; Hackett, 2005; Katz & Martin, 1995; Shrum et al., 2007; Wagner, 2004). Building on De Solla Price's perspective on big science, the studies of collaboration in science often use quantitative analysis, for instance, by looking at the number of authors of a publication as sign of collaboration. These quantitative studies indicate an increase of collaboration, but leave reasons for increase and the precise character of the collaborations unstudied. The more fundamental questions of 'what is collaboration?' and 'why collaborate?' are quite difficult to answer: "These deceptively simple questions have elicited and qualified answers" (Hackett, 2005: 668). Recent studies have tried to come to terms with these basic questions.

Starting with defining scientific collaboration, the notion of 'co-laboring' can be seen as the literal roots of collaboration (Maienschein, 1993). However, the crux of defining collaboration is defining what co-laboring actually entails as it can be more or less narrowly defined. For instance, Griesemer and Gerson first define scientific collaboration quite narrowly as "the work of teams of scientists with shared goals, such as formulating or testing particular empirical hypotheses, and with shared products, such as co-authored papers" (Griesemer & Gerson, 1993: 185). In addition, they suggest a wider view of scientific collaboration by presenting the Museum of Vertebrate Zoology as a collaborative effort: "an institution for conducting 'big' science – work that involves coordinating many people and substantial resources for long periods of time" (p. 202).

Maienschein agrees with this broader definition of collaboration: "The individuals should come together in pursuit of a shared goal. This may be true even if they do not articulate their goal in precisely the same way. Or indeed even if they don't articulate it at all" (Maienschein, 1993: 168). However, she distinguishes different types of collaborators consisting of primary collaborators – who share full responsibility for the project and its goals – and secondary collaborators who are less involved, like technicians or (data) collectors that only share responsibility for parts of the project.[12] In this context Shrum et

---

[12] Maienschein also notes that the distribution of credits among the collaborators is closely tied to this 'hierarchy of workers' and becomes manifest in the authoring of publications that result from

al. (2007) make a division between cooperation and collaboration, whereby in the first case people take each other's interest into account, while collaboration also involves a shared objective.

Consequently, Katz and Martin (1997) suggest a division between strong and weak collaboration. They refine the concept of collaboration by recognising different forms of collaboration, ranging from inter-individual collaboration to international collaboration. They distinguish different levels of collaboration from individuals to nations and collaboration can occur either between or within these different levels, using, respectively, the prefixes *inter* and *intra* (Table 1). "Thus *inter*national collaboration means collaboration between nations, while *intra*national collaboration means collaboration within a single nation" (Katz & Martin, 1997: 10).

	Intra	inter
Individual	-	Between individuals
Group	Between individuals in same research group	Between groups (e.g., in same department)
Department	Between individuals or groups in same department	Between departments (in same institution)
Institution	Between individuals or departments in same institution	Between institutions
Sector	Between institutions in same sector	Between institutions in different sectors
Nation	Between institutions in same nation	Between institutions in different countries

Table 1: different levels of collaboration and distinction between *inter* and *intra* forms

When dealing only with one of the prefixes a collaboration is called homogeneous, while a combination of intra- and inter- makes a heterogeneous collaboration. In addition, Katz and Martin pay attention to the boundary of collaboration, which they find highly dependent on social conventions and often "very 'fuzzy' or ill-defined" (idem: 8) and open to negotiation.

In a recent study, *Structures of scienctific collaboration* (Shrum et al., 2007), the authors further investigate the phenomenon of scientific collaboration through

collaboration. For instance, it is exemplified by co-authorship, the position of collaborators in the line of authors and the exclusion of collaborators as author.

a combination of quantitative and qualitative research on sixty-one scientific collaborations across different scientific disciplines, ranging from physics to space science, astronomy and material science. Although they looked at medical physics, they did not include the life sciences. The book distinguishes four different types of collaboration: bureaucratic collaboration, leaderless collaboration, non-specialised collaboration and participatory collaboration. Interestingly, they conclude that these organisational structures of collaboration are not specifically attached to certain types of science as they can be found throughout the spectrum of scientific disciplines studied. Except for particle physics, which is predominantly participatory as the collaborations have an 'Athenian-style democracy' (p. 115).

Next to defining scientific collaboration and its structure, the identification of factors that encourage the formation of research collaborations is an important subject in qualitative studies on scientific collaboration (Hackett, 2005; Katz & Martin, 1995; Shrum et al., 2007). Reasons for collaboration vary. As in big science studies, escalating costs of the development of large instruments are often put forward as a reason to collaborate. However, specialisation and multi-disciplinary research can also be an incentive to collaborate, as well as decreasing costs of travel and communication and the credibility of research. In addition, collaboration can be stimulated by funding organisations or it can have political motivations. More specifically, Katz and Martin (1997) suggest a difference between theoretical and experimental work, with experimentalists collaborating more than theoreticians. Also a division between basic and applied research can be made, with applied research tending to be more interdisciplinary as it requires a wider variety of skills or research partners. However, this is frequently argued the other way around as well: "the more basic the field, the greater the proportion of international co-authorships" (p. 4). Consequently, in many instances the reasons for collaborating are hardly straightforward, and they are difficult to pin down.

These recent studies of scientific collaboration primarily address collaboration in classic big science fields, but they can also be seen as attempts to give an overview of the phenomenon of scientific collaboration and create some order to build theory. Although the studies are not univocal, some important similarities in their results can be noticed. First of all, they identify increasing collaboration in science together with increasing reflection on the subject of which they are symptomatic themselves as well. They acknowledge the complexity of the phenomenon and the relative lack of qualitative study of scientific collaboration and they put forward different dimensions as an approach to study collaboration from a qualitative perspective. While the dimensions are formulated differently and subsequently address several different issues, they also cover similar features like the magnitude or extent of collaborations, reasons for collaboration (formation or purpose), the role of technologies, organisational aspects and typologies and the internal working of collaborations. Finally – and in light of the contemporary increase of collaboration – the various studies of collaboration point towards the importance of a critical approach towards scientific collaboration. While collaboration is sometimes seen as valuable in its own right, the various studies show that it also comes at a price.

Studying a multi-faceted phenomenon

This chapter has investigated different views on transformations in science and the increase of collaboration. I argued that scientific collaboration is a multi-faceted phenomenon. If there is not a single definition of collaboration, nor is there a single structure for collaboration. Collaboration takes place in various scientific fields and for various reasons, and there are also different approaches for studying scientific collaboration. Although the big science concept was first coined in order to come to terms with increasing dimensions in science, in time new approaches to transformations and collaboration in science have developed. Consequently, authors sometimes openly disapprove of the big science concept and decide to avoid it. As Shrum et al. (2007) have claimed:

> [Big science] is more than a label but less than a concept (…) the term has become a fen of vagueness and ambiguity through overuse (…) our conclusion is that the notion is simply too broad to be analytically useful. (p. 2)

Also scholars who are analysing the Human Genome Project and transformations in biology employ other perspectives (Balmer, 1996a; Balmer, 1996b; Cook-Deegan, 1995; Hilgartner, 1995; Glasner, 2002; Glasner, 2004; Sloan et al., 2000). Discarding the big science concept cleans the record and subsequently creates room for making a fresh start. However, does the history of a concept and its various connotations only provide unwanted information? Will the quality of the analysis improve when new concepts without any cultural baggage are used?

I will study scientific collaboration in biology as a multi-faceted phenomenon, building on the divergent approaches for studying collaboration, including big science. I found that the richness of the big science concept does not necessarily have to be a weakness – it can be a strength too. The big science concept acquired different meanings and has become a broad concept that covers various aspects of large-scale research. Most importantly, the big science concept provides a historical and cultural context for current transformations in science. Moreover, big science adds to more recent conceptualisations of transformations in science as it explicitly addresses issues of scale and the increase of scientific collaboration. In assuming that bigness is not pre-given but has to be created within the order of science, the study of the supersizing of science enlightens the ways in which collaboration in science is realized.

CHAPTER 2

Big Biology
Collaboration in the life sciences

In his *Reflections on Big Science* Weinberg (1967) already predicts biology to become the next big science:

> We are, or ought to be, entering an age of biomedical science and biomedical
> technology that could rival in magnitude and richness the present age of phys-
> ical science and physical technology. Whether we shall indeed enter this age
> will depend upon the attitude toward Big Biology adopted by biomedical sci-
> entists and government agencies that support biology. (p. 101)

Later he explains that his thoughts on big biology were inspired by Norman
Anderson, a 'biologist-cum-engineer' who in 1967 proposed the Molecular
Anatomy programme, aimed at cataloguing and characterizing all human proteins (Weinberg, 1999). Although the Human Genome Project preceded the
cataloguing of proteins, in 2001 the Human Proteome Organisation was
launched to foster international initiatives to investigate human proteins, currently known as 'proteomics'.

In this chapter I will explore whether biology can be considered big science. My consideration of the different meanings of big science has revealed
that from an empirical perspective this question can be answered in a quantitative and qualitative way. Increasing numbers seem to support the bigness of
biology. Most prominently, investments in research go up. In the light of
developments in molecular biology and genetics, the government of the United
States doubled the budget of its National Institutes of Health at the start of the
twenty-first century. Now the US spends $28.6 billion on federal research into

health, compared to $12 billion in 1998.[13] This is complemented by over $5 billion for non-health related biological research distributed by the National Science Foundation, which roughly doubled since the 1970s (see the analysis of the R&D Budget of AAAS).[14]

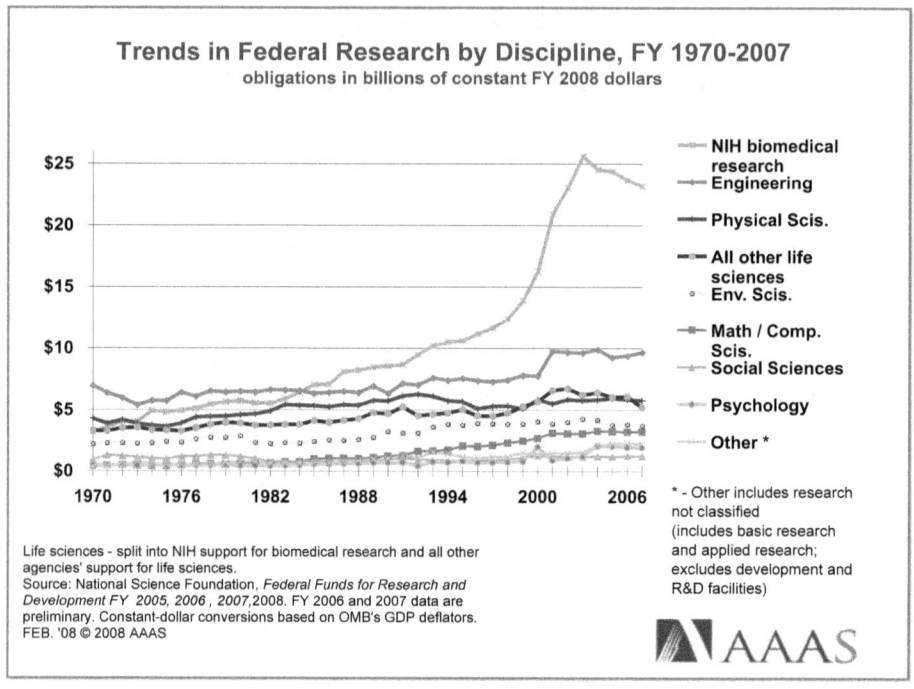

In the United Kingdom these numbers are mirrored by the budget for biomedical research going up to £550 million in the Medical Research Council budget.[15] Recent government plans even intend to merge this budget with the budget of the National Health Services, creating a new agency with a budget of at least £1 billion to support all biomedical research, from basic studies to clinical trials, resembling the US National Institutes of Health. In addition, the

---

[13] These numbers are distracted from R&D Budget and Policy Program of American Association for the Advancement of Science (AAAS), which continually analyses federal research expenditures: http://www.aaas.org/spp/rd/.

[14] Image courtesy of the former director of the R7D Budget and Policy Program Kei Koizumi.

[15] The UK numbers are distracted from the official websites of the councils and an article in *Nature* about Britain's budget plans for science *Brown's budget briefing*, published online 29 March 2006: http://www.nature.com/news/2006/060327/pf/440581a_pf.html.

Biotechnology and Biological Science Research Council, established in 1994, takes care of non-health related biology research with an annual budget of £336 million. In turn, the Netherlands are investing €571 million over a ten-year period in a newly established genomics research foundation, called Netherlands Genomics Initiative, in order to put the country on the global genomics research map. The European Union dedicated €2.25 billion to the new theme 'life sciences, genomics and biotechnology for health' in its 6th framework programme (2002-2006).[16]

Although it would be very interesting to see a more thorough analysis of increasing numbers in biology – one that not only looks at research investments but also considers the increase of manpower, publications, co-authorship, patents, research and educational programmes – in this thesis I explore the meaning of big biology by drawing on the qualitative empirical tradition of science studies.[17] As shown in the previous chapter, physics is the favourite subject of studies on big science and scientific collaboration, not biology. Moreover, many social scientists who study biology are primarily interested in researching the social, ethical and legal implications of new genetics research instead of organisational transformations. In a paper on the Human Genome Project, Alice Dreger (2000) even argues that the budget for ELSI research that accompanied the Genome Project "has tended to keep analysis focused on the potential applications of genetic research, rather than on the nature, meaning, and propriety of the research itself" (p. 171). In contrast, I will explore the current trend towards scientific collaboration in biology.

In order to characterise big biology I will compare current developments in biology to traditional big physics and show how big biology differs from big physics as a form of big science. First, I explore the history of collaboration in biology, demonstrating that next to the famous big physics, biology has a history of collaboration as well. Secondly, I will further investigate the claim that biology is not big science because of the absence of instruments as large as particle accelerators. Based on a consideration of the role of technology in

---

[16] The European figures can be found at the official website of the EU. Retrieved December 7, 2006 from http://europa.eu.int/scadplus/leg/en/lvb/i23012.htm. The current 7th framework programme (2007-2011) dedicates € 6100 million to health and €1935 million to food, agriculture, fisheries and biotechnology. Retrieved August 29, 2008 from http://cordis.europa.eu/fp7/-budget_en.html.

[17] Quantitative approaches already show increased collaboration in the life sciences, although emphasis lies on biotechnology and collaboration between academia and industry (Blumenthal, 2003; Blumenthal et al., 1996; Owen-Smith et al. 2002; Porter Liebeskind et al., 1995; Powell et al. 2005; Kleinman & Vallas, 2006; Zucker et al., 2001). I will explicitly address collaboration between science and industry in Chapter 5.

biology, I will argue that technology development in biology does cause organisational transformations, although they are less centrifugal than in big physics and have a more networked character. Thirdly, I will reflect on the way in which big biology is embedded in society, showing how contemporary biology is expanding in a societal context that differs from that of traditional big physics. This qualitative empirical analysis of the phenomenon of big biology finally leads to the characterisation of big biology as a contemporary networked form of big science.

A history of scientific collaboration in biology

While physics is commonly known as a collaborative science, biology is only recently portrayed as such, with the Human Genome Project functioning as a watershed. Recent narratives on biology often claim an unprecedented increase of collaboration in the life sciences. Does this imply that collaboration in biology is a recent phenomenon? When looking at histories of biology, earlier forms of collaboration can be identified. Before elaborating on the recent increase of collaboration, this section will therefore put collaboration in biology into a historical perspective. As such it builds on studies that have looked into particular forms of collaboration in the history of biology.

The origin of scientific collaboration

Scientific collaboration in biology is not new. To be more precise, natural scientists have been part of the first forms of scientific collaboration that have been described as the grand alliance between science and exploration in the 17th century (Capshew & Rader, 1992; Fernandez-Armesto, 2006). Explorations of the world did not only propel the mapping of earth and sky; they also widened our knowledge of the living world. Natural scientists joined expeditions into the unknown to collect new species (Magner, 1994). Naturalists had a passion for the accumulation and classification of facts about plants, animals and people from all over the world. When looking at maps and other representations of the travels into the unknown, you see beautiful flowers, colourful birds and indigenous people. This is nicely illustrated by a painting of Mount Chimborazo in South America that pictures local animals, plants and expedition members, including a small figure picking a flower, which is most probably the German naturalist Alexander von Humboldt.

Mount Chimborazo as pictured in the book: *Vues des cordillières et monumens des peuples indigènes de l'Amérique* (1810) by Alexander von Humboldt and Aimé Bonpland, p. 206. [18]

Until 1600 only about 6000 species of plants were known while in the year 1700 botanists had added 12,000 new ones, with similar accumulations in zoology. This not only further developed classification schemes – leading to Linnaeus *Systema Naturae* (1735)[19] and the evolution theories of Lamarck (1809/1984) and Darwin (1859) – but also caused some organisational changes in the acquisition of biological material. Infrastructural developments for transporting people and information were crucial for this early form of collaborative biology.

The early scientific expeditions gradually evolved into more coordinated multi-disciplinary research programmes that initially took the form of thematic years or decades, starting with the International Polar Year (IPY). Taking place in 1882-1883 and in 1932–1933 the first and second IPY's concentrated international research efforts geared to investigating the North and South Poles. Afterwards, the IPY's became a model for collaboration: "The experience gained by scientists and governments in international cooperation set the stage

---

[18] Image retrieved 14 February 2007 from: http://www.humboldt-portal.de/cd/Voelker_Amerikas/Abb_25.jpg
[19] Systema Naturae." Encyclopædia Britannica. 2008. Encyclopædia Britannica Online. Retrieved 25 August 2008 from http://www.britannica.com/EBchecked/topic/579163/Systema-Naturae.

for other international scientific collaboration" (International Polar Year, 2005). More concretely, the success of the polar years led to the organisation of the International Geophysical Year (1957-58)[20] – involving scientists from sixty-eight countries – and, later, the International Biological Program (IBP, 1968-1974).

> The International Geophysical Year, the modern descendant of the Polar Year, was a great success. The biologists, not to be outdone by the geophysicists, then proposed a biological "equivalent", the International Biological Program. The IBP is now in full swing, and it caused many nations to learn how to work together in scientific research with the highly practical purpose of improving the life of humankind. (Dunbar, 1971: 162)

In line with these thematic collaborations, the promotion of multi-disciplinary research in research programmes became a common addition to individual scholarships used by national and international funding agencies to distribute research money (interview Verschoor, 2007).

The International Biological Programme set the stage for large-scale research in biology (Kwa, 1987; Pirie, 1967; Dunbar, 1971). The original idea arose in Europe in the late 1950s and was advanced by the International Union of Biological Sciences, receiving important support from the US government. Although according to the original conception genetics and human population studies would be part of the core of the programme, in the light of upcoming environmental concerns systems ecology with its promise to control nature became the central issue of the programme. Consequently, the IBP can be seen as the first time in which ecology became big science, and since then it has also been analysed as such (Bocking, 1997; Parker, 2006; Schloegel and Rader, 2005).

Reasons for collaboration in biology

Historian of biology Jane Maienschein (1993) identifies three reasons for collaboration in the history of biology, related to research material, credibility and politics. As can be seen in the early history of biology, needing more hands to collect research material is a major incentive for collaboration in biology: "This may simply be a matter of needing more hands doing the same kind of work, or it may involve bringing together specialists who provide different types of

---

[20] *The International Geophysical Year*, The National Academies. Retrieved June 21, 2007, from http://www7.nationalacademies.org/archives/igyhistory.html.

expertise" (Maienschein, 1993: 167). Besides the so-called 'more-heads-(and-hands)-are-better-than-one factor', government, other funding sources and the larger scientific community have played an important role in the scaling-up of biology. Maienschein specifically identifies the 'credibility factor' and the 'political factor' that have influenced collaboration in biology. Scientific collaboration can lead to greater credibility because all researchers bring their own credentials and acceptability from their own research community into the collaboration. In addition, cooperation has been a means for biology to become more visible in the light of physics and chemistry, the more flourishing sciences in the middle of the 20th century. Maienschein thereby implies a relation between the scale of scientific practice and the significance attributed to that science.

Finally, Maienschein distinguishes political reasons for collaboration. She argues that the ideal of collaboration – as held by government and foundations – was an important incentive for collaboration in life sciences research. Especially in the United States research foundations and the government played an important role in the move towards collaboration:

> It was precisely the availability of new money that ultimately created the push to collaboration and cooperation in the 1920s (...) The foundations, and the government agencies that began to emerge at about the same time, pushed away from individual research efforts and toward the development of a "collective, communitarian character" for science. The foundations purposefully wished to promote cooperation. (Maienschein, 1993:181-182)

As a result scientists started to work together or at least they appeared to be working cooperatively. Moreover, an increased tendency to collaborate in biology, it is argued, has played an important role in defining and redefining of biology. In other words, the character of biology as a science is also formed through its collaborations.

Large-scale biomedical research

Among historians of biology an important debate concerns the emergence of large biomedical complexes (Rasmussen, 2002). A common view is that they first appeared in World War II together with the large-scale projects in physics organised by the government of the United States. Large-scale research projects in the biomedical realm focussed for instance on the development of penicillin and blood products (Creager, 1998; Neushul, 1993). However, what is argued in the case of physics is argued for biology as well: collaborations with industry already started before World War II (Capshew & Rader, 1992; Rasmussen, 2002; Seidel, 1992). On the basis of research into industrial collaborations before the war Rasmussen argues: "'Big' biology, performed in comparatively large groups with substantial budgets, was already commonplace by 1941, financed by drug and chemical companies or, in selected fields less favoured by industry, by philanthropies such as the Rockefeller Foundation" (Rasmussen, 2002: 116).

The academic-industrial collaborations started in the 1920s and 1930s when American pharmaceutical firms invested in research as a competitive strategy in combination with medical reform movements aiming to make science the basis of therapeutic practice (Rasmussen, 2002). The firms opened in-house laboratories, but also turned to university scientists as source of expertise. "In a typical collaboration, a firm would fund an academic researcher and stipulate that new processes and inventions be patented and assigned, or licensed on favourable terms, to the firm. Royalties to the university hosting the research would be part of the arrangement" (p. 120). Moreover, Rasmussen argues that these arrangements set the stage for large-scale research arrangements during World War II. Although the number of collaborations in the biomedical sciences increased during the war, only afterwards the government became a leading patron of the expanding life sciences – funded by the National Science Foundation and the National Institutes of Health – making industrial sponsorship unnecessary until the Reagan era brought industry back into the picture.

In addition, the first forms of large-scale physics research which emerged in the first decades of the twentieth century gave rise to radiobiology (Creager & Santesmases, 2006; Lenoir & Hays, 2000; Schloegel & Rader, 2005; Westwick, 2003). This connection between large-scale physics and biology became even stronger after World War II when physics laboratories had to reposition themselves, as investments in physics became controversial, first because of the devastating power of the Atomic Bomb and later because of the debate on Nuclear Energy. As a result, the national laboratories in the United States started to incorporate life research in their strategy, as this brought clear societal benefits and also was in line with the increasing attention for environmental pollution. In this context, the large-scale research organisations for physics gave a new impetus to biology research, which became particularly apparent in radiobiology and environmental research (Bocking, 1997; Creager & Santesmases, 2006; Schloegel and Rader, 2005). Moreover, in the context of their interest in the effects of radiation on the human genetic make-up, it was the Department of Energy that first envisioned large-scale research collaboration in molecular biology. This department used its experience with organising large-scale physics research in shaping the Human Genome Project, which opened the way for other large-scale research projects in the context of genomics and post-genomics research.

The emergence of molecular biology

The emergence of molecular biology in the second half of the twentieth century has been crucial in recent developments towards collaboration in biology (De Chadevarian, 2002; Magner, 1994; Strasser, 2003a; Strasser, 2003b). Initially, molecular biology was not seen as a collaborative field of study. Molecular biologists were even hostile to the International Biological Program – one of the first large-scale research projects in ecology – because they were afraid it would drain off some of the flow of funds to molecular biology (Kwa, 1987). In *Epistemic Cultures* (1999) Karin Knorr-Cetina argues that unlike in big physics, cooperation in molecular biology is 'impossible' as it is an 'individual, bodily lab-bench science'. The so-called 'collaboration-problem' can be explained by the 'individual ontology' of molecular: the individual scientist working on his or her own project is the central unit in the research practice of molecular biology, next to the individual laboratory represented by the laboratory leader. Knorr-Cetina argues that even collaboration in biology takes place within this 'individual ontology', following a 'logic of exchange': "Participants in the lab render services to each other in exchange for other services and co-authorship. Equally, the leader is often approached with requests for services by other labs, which he or she gladly renders, knowing that return favors will be forthcoming and might be needed" (Knorr-Cetina, 1999: 255).

Nevertheless, molecular biology has been a major force in the recent development towards collaboration in biology. In Europe the creation of Centre Europeen de Recherche Nucleaire (CERN) in 1953 and the establishment of the European Space Research Organisation in 1962 was followed by the creation of the European Molecular Biology Organization (EMBO) in 1964 and the European Molecular Biology Laboratory (EMBL) in 1974 (Strasser, 2003a). EMBO was originally an idea of Leo Szilard, a nuclear physicist turned molecular biologist, who envisioned a international laboratory for molecular or fundamental biology patterned on the CERN model (Strasser, 2003a). In 1963 a group of molecular biologists discussed the possibilities during a meeting in Italy and decided that a European organisation was more realistic than an international one. At that time many organisations were pushing international collaboration in the life sciences and the creation of an international laboratory:

> We tend to forget that in the 1960s, EMBO had to face many competing projects to develop international cooperation in the life sciences at the European,

Atlantic and global level. These projects were backed by well-established organizations such as the Council of Europe, EURATOM, the North Atlantic Treaty Organisation (NATO), the Organisation for Economic Cooperation and Development (OECD), the United Nations Educational, Scientific and Cultural Organization (UNESCO) and the World Health Organization (WHO). (Strasser, 2003a: 542)

However, the architects of EMBO have been able to tactfully manoeuvre the political waters and establish their organisation in the context of European integration. Moreover, 'molecular biology' became the new focus of European science policy.

In the context of molecular biology, global collaborations took shape as well in the form of mapping and sequencing projects (García-Sancho, 2008; Gaudillière & Rheinberger, 2004; Kohler, 1994; Sulston & Ferry, 2002). Alliances around the model systems Drosophila and C. Elegans particularly catch the eye. Studying respectively the fly and the worm scientists began to form research networks that exchanged research material, information and divided work over different laboratories. The type of materials they exchanged gradually transformed (García-Sancho, 2008). The communities began sharing mutants and know-how and in time scientists substituted the exchange of specimens by increasingly abstract representations of them. Also the development of technologies became part of the joint efforts to map and sequence the organisms. These projects have been forerunners of the Human Genome Project. In particular the collective work on the C. Elegans set the stage for the collaborative work on the human genome, as its meetings overlapped at the very beginning with what later became the Human Genome Project (Sulston & Ferry, 2002).

A new era of collaboration in biology

Although views on the character and the use of the Human Genome Project widely vary (Balmer, 1996a; Balmer, 1996b; Cantor, 1990; Glasner, 1996; Glasner, 2002; Hilgartner, 1995; Hull & Ruse, 1998; Kevles & Hood, 1992; Sulston & Ferry, 2002), the project is often presented as a turning point in biology. The Human Genome Project not only symbolises a new scientific approach but also a new way of organising research. On the website of the National Institutes of Health the Human Genome Project is presented as "the natural culmination of the history of genetics research". [21] Think for example of the discovery of the

---

[21] 'All About The Human Genome Project', Retrieved February 20, 2006 from: www.nih.org

structure of DNA, the subsequent isolation of DNA in a test tube and the development of recombinant DNA and sequencing methods. Also organisationally, the Genome Project is pictured as an important step in the history of biology:

> In June 1986, at a Cold Spring Harbor meeting, Gilbert declared that the project would be vastly accelerated by putting several thousand people to work on it, estimating that, at a sequencing cost of one dollar per base pair, the complete human sequence could be obtained for three billion dollars. It would be a big project for biology. (Kevles, 1992: 22)

To indicate the magnitude of the actual project: the public consortium consisted of sixteen groups spread about forty-eight laboratories world-wide (Balmer, 1996a; Balmer, 1996b; Glasner, 2002). Each participating country established its own sequencing centres, with their own administrative structures and funding arrangements. With additional efforts in the private realm the publication of the completed maps of the human genome contained the names of some 520 scientists.[22]

When focussing on the character of the Human Genome Project, sociologist of science Peter Glasner (1996) identifies a transition from the traditional Mertonian scientific community to a new 'collaboratory' model. The term 'collaboratory' was coined by Wulf (1989). It refers to the way in which laboratories are working together over their respective walls in international collaborations supported by information and communication technologies, making the physical location of the laboratories irrelevant. This explicit and formalised network building that presumes certain public rules of membership involves a transition from the notion of the 'invisible' college to that of a very 'visible' college or network (Price, 1963; Crane, 1972). In turn, Glasner describes sequencing efforts – in opposition to more conventional big science arrangements – as "conducted by mainly small groups of scientists contributing to a clearly defined long-term goal" (Glasner, 1996: 109). A 'collaboratory' is then defined as "technology, tools and infrastructure that allow scientists to work with remote facilities (co-laboratory) and each other (collaborat-ory)" (Glasner, 1996: 111, citing Lederberg and Uncapher, 1989: 3). Or in other words: 'a laboratory without walls' that provides seamless access to colleagues, instruments, data, information and knowledge.

In addition to being a turning point, the Human Genome Project is presented as a starting point. It is portrayed as the foundation of major current

---

[22] For an analysis of the transformation towards multiple authorship, see also Galison, 2003.

research endeavours: "With the completion of the Human Genome Project, we are now entering the Genomic Era" (NIH, 2003). In policy plans of the Human Genome Research Institute the project quite literally serves as the foundation of future plans.

The future of genomics rests on the foundation of the Human Genome Project
(Collins & Green, 2003)[23]

In the slipstream of the Genome project various collaborations have been set-up. First of all, numerous sequencing projects target various kinds of life from the gigantic to the microscopic.[24] In addition to genomes, other parts and processes of cells are charted within the contexts of so-called '-omics' research, like proteomics and metabolomics and nutrigenomics. These collaborations are accompanied by a shift from biology as a data-poor to a data-rich science,

---

[23] Image courtesy of Francis Collins.
[24] For example, eighteen species are tackled by the NHGRI: ranging from the elephant to the hedgehog (Elias, 2004; Hopkin, 2004). In addition, the genes of various ancient mammals are reconstructed (Grimm, 2004; Khamsi, 2004; Pennisi, 2004). In turn, Craig Venter is now sequencing all kinds of microbes (Hopkin, 2005; Venter, 2005).

which asks for computational tools, integrated knowledge systems and interdisciplinary work (Glasner, 2002). It is argued that this research can only be performed through multi-disciplinary collaboration. The expertise needed to perform the research has to come from different disciplines and sub-disciplines that need to be combined. Moreover the importance of technology brings the need for engineers and computer scientists and the increasing importance of modelling asks for mathematicians. [25] Currently, the interaction of the different elements and processes of the living are emphasised in 'systems biology', which is envisioned to lead to collaborations that are even larger than the Human Genome Project. Or as phrased in an editorial on 'post-genomics cultures' in *Nature*: "Like it or not, big biology is here to stay" (Editorial *Nature*, 2001: 545).

As a result scientific work becomes explicitly pictured as collaborative. For example, the Erasmus Medical Center in Rotterdam, the Netherlands, says it is looking for 'teamplayers' on a DVD aimed at recruiting new scientists, entitled "Genomics is Teamwork" (Van Vliet, 2004). The scientific work is explicitly pictured as collaborative: "I think teamwork is the essence of modern science. (...) At least working in the field of biology or medicine, it's impossible to do the trick now – all kind of specialist tricks – by yourself", as Professor of Epidemiology Cock van Duijn explains. Other scientists confirm Van Duijn's view on the importance of collaboration. [26]

> I think the biomics core facility [a genomics, proteomics and bioinformatics technology service facility and tissue bank] [27] also has been a glue for different departments who used to work on their own islands and now they are brought together in this facility to work together (...) Our expression profiling study would not have been possible without this cooperation. The study was done in-house, entirely in-house, with new technology, but it depended on cooperation between the department of haematology, pathology, genetics, bioinformatics, to name a few. And this is a key element of our future progress.

The DVD presents collaboration in contemporary life sciences research as a requirement and also stresses the important role of new technologies in collaboration.

---

[25] See also Jamie Lewis' PhD thesis (2009) on the proteomics initiative 'Computing genomic science: bioinformatics and standardisation in proteomics' (Cesagen project on proteomics) and Bart Penders' PhD thesis (2008) on international nutrigenomics research collaborations.

[26] The researchers that appear in the DVD are professors Cock van Duijn, epidemiology; Frank Grosveld, cell biology; Bob Löwenberg, hematology; Theo Luider, proteomics; Ab Osterhaus, virology; André Uitterlinden, internal medicine; and Peter van der Spek, bioinformatics.

[27] For more information on the biomocis core facility see http://www.erasmusmc.nl/-biomics/index2.html. Retrieved February 14, 2007.

In addition, biology laboratories are built to facilitate collaboration. Robert Venturi designed the Lewis Thomas Molecular Biology Laboratory (1986) at Princeton University especially with collaboration in mind as "microbiologists are social people who feel strong ties to their scientific community" (Brand, 1994: 179). According to the department chairman at that time, Dr Arnold Levine, the highest priority in planning the building was to force interaction between research groups and individuals. This design became a model for several other labs. The James H. Clark Center at Stanford is a more recent example of architectural materialisation of collaboration in biology (Check & Castellani, 2004; Hall, 2003; Hubbard, 2003; Shwartz, 2003). Stanford University presents the building as "the centerpiece of Stanford University's Bio-X program – an innovative campuswide initiative designed to foster interdisciplinary research in the biosciences by bridging the worlds of biology, medicine, engineering and the physical sciences".

James H. Clark Center [28]

---

[28] Photograph by Nigel Young, courtesy of Foster + Partners.

The building was completed in 2003 and according to its major sponsor, James Clark, it is a truly collaborative environment: "This is not a place were you come and sit and sort of pan out your things all by yourself. This is an environment where you come to interact with others, and more important to get your students to interact with other students" (Clark quoted in Hubbard, 2003). Next to multiple spaces that encourage intermingling, the building materialises cooperation through its open mega-labs: "Cauldrons of creativity where researchers from different disciplines have been willingly thrown together in the hope that close encounters will spark undreamed of discoveries in biology and medicine".[29] The labs have a thematic organisation and the set-up of the building is flexible: "The furniture is on wheels. Laboratory benches are hooked up to exposed utility systems hanging from the ceiling, easy to get to in case someone wants to push a wall or two out of the way" (Hall, 2003). Chemist Tom Wandless – who used to work in 'a very monastic atmosphere of solitary research' and is now finding himself next to biologists, computer scientists and engineers – calls it an experiment in "social engineering". An experiment that seems to be successful, as already a year after its opening "collaborative projects were brewing", according to an article on developments within the centre (Baker, 2004).[30]

Collective ways of knowing in biology

The dispersed nature of biological research material has been an important incentive for collaboration, as is true of recent developments in molecular biology. In addition, science funding, science policy and the shaping of biology vis-à-vis other scientific disciplines have contributed to increasing scales in the life sciences. This intricate relationship between research material, research approach and socio-political factors is the main concern of science historian John Pickstone (1993; 2000; 2007). With the historiography of medicine as starting point, Pickstone constructed a model of the development of science,

---

[29] About the James H. Clark Center. Retrieved November 20, 2006 from, http://www.stanford.edu/home/welcome/campus/clark.html
[30] Similar developments can be seen in Europe. For instance, the new building of the Max Planck Institute of Molecular Cell Biology and Genetics in Dresden opened its doors in 2002 (Max Planck Institute of Molecular Cell Biology and Genetics, Dresden: 'The building. Communication through architecture.' Retrieved February 14, 2007 from http://www.mpi-cbg.de/research/profile.html). In addition, the Manchester Interdisciplinary Biocentre has also materialised its mission 'to promote interdisciplinary, challenge-oriented biosciences and biotechnology' in a new building (interview McCarthy, 2005; Manchester Interdisciplinary Biocentre, University of Manchester: 'The Building'. Retrieved February 14, 2007 from http://www.mib.ac.uk).

technology and medicine over time. He distinguishes four specific forms of scientific work with corresponding social relations, nested in particular institutional and socio-political environments. These four 'ways of knowing' are, respectively, natural history, analysis, experimentalism and technoscience. When looking at collaboration in biology from this perspective, it becomes clear that collaboration has to be seen as part of particular ways of knowing.

For instance, early collaboration in biology to assemble research material is clearly part of Pickstone's natural history model. In this way of knowing the description and classification of objects or systems is key: "From the seventeenth century, at least, natural history has been the study of 'what we have' – in databanks, or public or private collections" (Pickstone, 2000: 11). This particular kind of scientific work has to be placed in the context of the Renaissance, trading nations, empire building, cabinets of curiosity and the establishment of scientific societies and national museums. In contrast, beginning in the nineteenth century, the 'Age of Analysis' gave rise to an analytical shift, which took place in interaction with the emergence of a mechanical view, processes of rationalisation, the formation of disciplines and the development of research instruments like the microscope. In the late nineteenth century experimentalism and invention became central as means to control nature and create novelty. Although this new experimental work in biology still encompassed collection, observation and analysis, its proponents often portrayed themselves in opposition to older naturalists which they dismissed as 'stamp-collectors' (Pickstone, 2000; Maienschein, 1991). It is in the context of analytical and experimental biology that we can recognise the emergence of the laboratory and the individual mode of working in biology as described by Knorr-Cetina (1997).

However, the historiography of biology also shows how the analytical and experimental way of knowing in biology gradually acquired a more collaborative character. In these ways of knowing a shift from individual to collaborative work has taken place, in interaction with the development of molecular biology and genomics research. In addition, forms of collaboration between laboratories, government and industry appeared in biology at the beginning of the twentieth century (Ceager, 1999; Kwa, 1987; Neushel, 1993; Rasmussen, 2002; De Chadevarian, 2002; Strasser, 2003a; Sulston & Ferry, 2002). Connections were first built in the context of biomedical sciences before and during World War II. These developments comply with what Pickstone calls 'technoscience':

> *Techno-science*, including techno-medicine, I take as first created at the end of the nineteenth century, when certain laboratory products (or processes) became commodities, so partially reconstructing the social relations of STM to include industrial research laboratories, linked to universities, to state laboratories and

to other institutions, in ways which have since become more and more areas of STM. (Pickstone, 1993: 434)

This collaborative way of knowing is also described in recent theories of science and innovation.

New socio-technical arrangements in biology

Literature on big science and scientific collaboration often presents technology as a major incentive to collaborate. In big physics large instruments are seen as the most important reason to centralise, because not one university or even one country can afford to build and maintain these large instruments on its own. Nevertheless, important differences can be noted between big physics and big biology. For instance, reasons for collaboration in biology have been found in the dispersed character of the research material, in a changing research approach and in the political stimulation of scientific collaboration. This is why scholars, when comparing collaboration in biology to big physics, wonder in particular about the role of technology in biology. Can biology be considered big science while instruments are significantly smaller than the technologies of big physics? This section investigates the role of technologies in the recent increase of collaboration in the life sciences.

With investigations into the molecular level, vast changes have occurred at the instrumental front of biology. Not only special instruments developed, but also information and communication technologies became integrated in the research process:

> For life at the molecular level is only knowable through complex and expensive apparatus: electron microscopes, ultracentrifuges, electrophoresis, spectroscopy, x-ray diffraction, isotopes and scintillation counters and their links with the information-processing capacities of computers, and now, with the information dissemination capacities of the Internet. (Rose, 2001: 15)

Not surprisingly, technology takes a central place in views on the origin of the Human Genome Project. On the one hand it is pictured as a centralising force: "The human genome project was borne of technology, grew into a science bureaucracy in the U.S. and throughout the world and is now being transformed into a hybrid academic and commercial enterprise".[31] On the other hand, it is clearly pointed out that the project did not start with tools alone, but with an idea of a person who was embedded in a specific organisational and cultural context:

> This human genome program (HGP) has an unusual origin. It was not initiated by a committee of molecular geneticists dealing with a pressing need or by the major biomedical funding agency, the National Institutes of Health (NIH). Instead, it was advanced by a politically astute administrator [Charles DeLisi] in the Department of Energy (DOE), convinced that the powerful tools of molecular development made it appropriate to introduce centrally administered "big science" into biomedical research. (Davis et. al., 1990: 342)

This second perspective takes the complex interaction between technological and social developments into account (Blume, 1992; Bijker & Law, 1992; Clarke & Fujimura, 1992; Disco & Van de Meulen, 1998; Misa, 2004).

Building on this last approach and empirical research, in this section I investigate different ways in which technology development interacts with organisational changes in life sciences research, focussing on aggregation, centralisation and networking. First, the expansion of research arrangements will be exposed through the development of one particular instrument that played an important role in research at the molecular level: Nuclear Magnetic Resonance Spectroscopy. Secondly, I will consider the establishment of a Technology Facility at the Biology Department of York University and how it addresses centralisation around various technologies. Thirdly, based on the informational turn and the development of bioinformatics I will discuss the integration of information and communication technologies and the building of research networks in biology. Thus it will be revealed in which ways technology development is an important factor in the growth of biology, but I also explicitly argue that technologies are part of a process in which research arrangements are actively reshaped by people and organisations, underlining earlier critiques of the technological deterministic character of big science literature (Capshew & Rader, 1992; Westfall, 2003; Wyatt, 1998).

---

[31] See Cook-Deegan (1995) 'Origins of the Human Genome Project'. Retrieved February 2007 from http://www.fplc.edu/risk/vol5/spring/cookdeeg.htm

First of all, the development and use of larger instruments in biology can be noticed, leading to higher levels of aggregation in the organisation of research that resemble centralisation around large instruments in big physics, although on a smaller scale. The introduction of the Nuclear Magnetic Resonance Spectroscope[32] in biology shows the emergence of aggregation in biology research. Although NMR spectroscopy has its origins in the physical and chemical sciences, it became employed in biology to study molecular structures from the 1930s onwards (Zallen, 1992). The introduction of NMR spectroscopy in biology research was part of a broader effort of the Rockefeller Foundation to stimulate cross-disciplinary research. The integration of the NMR spectroscope came together with the further development of the instrument, increasing costs and increasing scales of research arrangements, from the local to the national and international level. Dr Rien de Bie —former research manager at the Bijvoet Centre for Biomolecular Research at Utrecht University in the Netherlands — witnessed these increasing scales of research arrangements around the instrument during his career (interview De Bie, 2006). [33]

De Bie recalls how the first NMR spectrometer for biological research entered the chemical department of Utrecht University in 1963 where it initially functioned in the local context. It was used in chemical research on biological structures, starting with the study of very small molecules but more became possible when the instrument improved. However, the costs of the instrument increased as well. To illustrate, the first NMR apparatus that De Bie bought had a price tag of 150,000 Dutch guilders (which is slightly less than 75,000 euros) while the last NMR he acquired cost 6 million euros, even though the last machine was probably 10,000 times better than the first "it indicates the enormous increase in scale" (interview De Bie, 2006). Nowadays the best NMR apparatus will cost about 12 to 15 million euros.

---

[32] The spectrometer has its origins in the physical and chemical sciences but has been appropriated by biology from the 1930s onwards (Zallen, 1992). In short, spectroscopy uses light and its absorption pattern to determine substances in a sample and it is used in biology to characterise interactions between molecules.

[33] The Bijvoet Center is a collaboration between Utrecht University and the Netherlands Foundation for Chemical Research (SON) devoted to structural biology. Information retrieved from the website of the Bijvoet Centre, Historical Background: http//:www.bijvoet-center.nl/about/-history on 30 November 2006.

A NMR Spectrocope[34]

In the 1980s there was a first move towards larger scales when a shift from local ad-hoc financing of research instruments to structural investments in a national context took place:

> It was during this period that in the Netherlands the idea developed that if you want to use expensive technologies – which means electron-microscopy, NMR spectroscopy, roentgen defraction, ultracentrifuges and that kind of thing – you have to start thinking seriously about how to go along (interview De Bie, 2006).

Based on existing practices in physics, this resulted in a policy of concentration and the emergence of so-called 'para-university research institutes'. These new institutes – geared in particular to acquire large investment funds on a national level – came just in time, as the costs for the technology could not be covered within the normal university financing system anymore: "they were dispropor-tional" (idem). In 1988 three national collaborative centres were established, including the Bijvoet Center for Biomolecular Research, devoted to structural biology.

---

[34] Image courtesy of the Bijvoet Centre, University of Utrecht: http://www.nmr.chem.uu-.nl/-bijvoet_brochure/nmr_750.gif..

The next scale increase was mainly policy induced as internationalisation became a priority in national research policy from the end of the 1980s onwards. Suddenly the Dutch government embraced 'internationalisation' as a priority, for instance through the development of ties with neighbouring countries in the so-called 'border regions'. De Bie managed to acquire 7,5 million guilders from the Dutch government, on the condition that his team would work on the positioning of the Netherlands within the international NMR research community: "So we said: if the Ministry wants internationalisation, we simply do it. And darned, we were successful" (interview De Bie, 2006). Next to national stimulation of internationalisation the Europeanisation of research arrangements started slowly. De Bie does not remember people in Utrecht being aware of the existence of the first and second Framework Programme. However, as no research infrastructures were in place on the biological side of chemistry, Utrecht decided to take part in the third Framework Programme together with centres in Florence and Frankfurt.

> We just said: we are the European consortium for European infrastructures for NMR. Quite arrogantly: we are the best, you know, that kind of story, and they believed us in Brussels. So Brussels said while looking at me: if you manage this network, we will take a close look at it and that made us quite a lot of money. And it always was like this. (...) So we were very successful, and I still do not understand completely why, but it looks like we already invented 'eurospeak' back then. (idem)

This led to the establishment of European collaborations for NMR spectroscopy that still exist today.[35]

Nowadays, the Bijvoet centre is still very successful, but De Bie also observes some serious problems concerning the use of NMR technologies for research, concentrating around the maintenance of technologies. Often, research funding is still geared to research projects or individual scientists, without taking into account that the technology has to be financed as well. Once the technologies are acquired with investment funds, they need housing, maintenance and technical specialists. These are mostly not provided, especially not by

---

[35] De Bie states that 'eurospeak' helps to write a successful European research proposal. He illustrated this with a story on the changing policy direction that came together with FP 5. Before, you could just assemble your research friends and draft a proposal for a research network, but at the launching meeting of FP 5 it became clear to De Bie that the new philosophy of the commission was not collaboration-based but problem-based: identify a societal problem, come up with a solution, figure who and what is needed for this solution, bring these people together and use this as fundament for your proposal. With this recipe in mind De Bie wrote five proposals of which four were rewarded: 80% success.

foundations for cancer or asthma who explicitly only want to fund research. This implies that the pressure on the normal staff becomes bigger, or the budget for staff members has to be traded against more technicians. The result is a work-overload for the group and its leader, who also have increasingly busy agendas as the research environment keeps expanding: "Compared to ten years ago, the amount of invitations that our top scientists receive to give a paper or take part in a conference is very high. Moreover, they are coming from more far-away countries like China, South Africa, Brazil, Australia, etcetera, which almost never happened earlier" (interview De Bie, 2006).

To conclude, the career of De Bie witnessed how instruments are actively integrated into the biological research practice by various actors.[36] Not only the instrument is constantly adapted – first gearing it towards biological research and making it more sophisticated and expensive ever since – but also research practices are transformed, expanding into larger scales. The development of the technology is integrated in a process in which research arrangements are actively reshaped by people and organisations. Moreover, it has become clear how this is a continuous development in which configurations are constantly reshaped. Next to increasing scales, new divisions of work become apparent. The management of the research process increasingly requires attention. De Bie turned himself into what is now called a research-manager, a new role that has been important to the success of the Bijvoet Centre. In addition, technicians become more important to develop and maintain the technologies. Nevertheless, the academic system does not fully accommodate such new roles yet.

Centralisation: The development of a Technology Facility

Next to aggregation around a single technology, a second organisational movement pertains to centralisation around several different technologies as recognised in the establishment of local central technology facilities. While researchers and laboratories always used to buy technologies individually, this strategy has its limits nowadays (interview Bowles, 2006; interview De Bie; Perkel, 2006). It is increasingly difficult to buy all the technologies needed when starting a lab. In addition, keeping the technology in an existing lab up-to-date and functional is almost undoable. This gives rise to several alternative practices, such as sharing or leasing laboratory equipment, the use of so-called 'kits', or black-boxed tests that perform already standardised research processes and

---

[36] For De Bie's reflections on the general research system, see his retirement speech (De Bie, 2004).

the use of (commercial) service centres as a form of outsourcing (Kleinman, 2003; Wolthuis, 2006). However, around the world increasingly the development of in-house Technology Centres can be recognised, where scientists can go to make use of specific technologies. The Technology Facility in the Biology Department at York University (UK) is an example of this recent development towards centralisation around the technological repertoire in biology.

The Technology Facility is part of the new Biology Department that opened in July 2003.[37] The facility is explicitly developed to function as a hub within the department where scientists can go to perform specific techniques. "The Technology Facility is a unique central support facility for bioscience research and benefits from over £6 Million of capital equipment, 20,000 m2 of purpose built laboratories and 16 expert staff members."[38] It gives access to various different technologies: imaging and cytometry, proteomics and analytical biochemistry, genomics, bioscience computing, molecular interactions and protein production. The Technology Facility is a great success; it improved the status of the Biology Department at York and is instrumental in the attraction of good scientists. Moreover, the centre propels new collaborations as scientists with different research aims, skills and backgrounds meet up through their need to use the same technique and thereby share their experiences and expertise.[39]

The central person behind the Technology Facility is Professor Dianna Bowles.[40] When she tells her story about the establishment of the facility she confirms the strong connection between recent transformations of technologies that are used in biology research and organisational changes. She witnessed these changes during her career and distinguishes three phases: from self-contained research labs via large-scale sequencing projects towards the current post-genomics era in which researchers simply need to incorporate new technologies in their research if they want to be part of the frontier of science. The Technology Facility is built in reaction to academic research falling behind the industrial capacity to innovate.

---

[37] I have interviewed several leading staff at York University about the development of the Technology Facility and the way in which it functions (interview Bowles, 2006; interview Hubbard, 2005; interview Sanders, 2005; interview Van der Woude, 2006).

[38] The website of the Technology Facility. Retrieved February 6, 2006, from http://www.york.ac.uk/biology/tf.

[39] According to Pillmoor, the director of the Technology Facility. Pillmore cited in *Bioscience Case Study* on University of York Technology Facility. Retrieved from the web, December 2006, http://www.bioscience-yorkshire.com/assets/oldassets/bio/c_York_University.pdf

[40] More detailed information on her scientific career can be found at the staff pages of the Biology Department of York University. Retrieved February 6, 2006 from: http://bioltfws1.york.ac.uk/biostaff/staffdetail.php?id=djb

Bowles explains how technological developments led to a shift from public to private domination on the frontiers of science in the 1980s, as companies invested money in specific technologies they needed, while universities were not able to acquire these technologies:

> So more and more became possible technically, but it only became possible with bigger and bigger capital facilities and items of equipment. (…) So then everything started to stall, because people just knew what could be done, but laboratory after laboratory, university after university, and institute after institute were increasingly falling behind because they did not have the big capital facilities and the know-how to make use of all these opportunities that were actually available to pursue their science. (interview Bowles, 2006)

This led to new funding initiatives in the public sector to increase capability at the UK universities, led by the Welcome Trust: "the Welcome Trust provided funds and thereby put pressure on the government to put in a lot of money in as well" (idem). This resulted in the Joint Infrastructure Fund (JIF) that provided money for the upgrade of public university facilities (Fitter, 2003).[41]

The Infrastructure Fund was fundamental to the establishment of the Technology Facility in York. The department received £21 million. "We prepared a bid for essentially a new Department of Biology where the heart of the department was the Technology Facility" (interview Bowles, 2006). The sharing of common technologies formed the central idea behind the plan. As with the development of genomics and post-genomics research, scientists increasingly use the same technologies, the facility needed to contain these so-called 'platform technologies' (Chang et al., 2000; Keating & Cambrosio, 2003).

> You can be interested in spiders or in plants or whatever, and they are all underpinned by the same technology platform. But individual research groups, do not need to use those technologies 100% of their time; they just jump in and jump out when their research dictates the need. So the best way to do it is actually to put all of these technologies in one place. (interview Bowles, 2006)

Moreover, Bowles explicitly aspired 'bigness', instead of asking for several smaller grants which is common practice in biology. She just asked scientists what kind of instruments they would need in the first ten years without consideration of costs and included these in the plan. This 'ambition of grandeur'

---

[41] The Joint Infrastructure Fund was established by the Wellcome Trust in cooperation with the Department of Trade and Industry and the Department for Education and Skills through the Higher Education Funding Council for England.

probably even explains the funding success, as York was one step ahead of other departments, which still focussed on procuring small funds.

Next to the sharing of technologies, the development of professional operational skills is fundamental to the facility. With the growing complexity of instruments the knowledge and skills required to use instruments has become more important and operational expertise crucial. Nowadays, it is often a full-time job to keep up with the latest technology developments and this requires specialists. Through the Infrastructure Fund, Bowles could exactly address this problem of technological expertise, as it also provided funds for technical staff.

> The way the public sector works, the public funds work, is that it is very difficult to actually fund what you call core technical personal who have high specialist expertise (...) and the key issue about that initiative – which is why I got very, very interested – is that for the first five years they allowed funding for the technologists. (interview Bowles, 2006)

At first sight, the increasing importance of technicians seems to be a classical case of work division as part of the industrialisation of science, but this is actually not the case. Bowles explicitly did not want the facility to be a service centre – "a bit like a gas station you could turn-up to and get your results back having no learning experience at all" (idem) – but a training site where scientists learn to use the technologies themselves. Consequently, researchers can now relatively easily incorporate the technologies into their research, allowing for further integration of technology in research processes and thereby inherently changing the character of the research process.

However, as at the Bijvoet Centre in the Netherlands, the problem of the financing of technical staff is quite acute in York. Initially, the technicians were paid by the government, but now other sources need to be found. Doing work for industry is one way for the facility to get self-funded, but this would mean using public investments to help business, instead of assisting and training public scientists. Consequently, Bowles emphasises that Research Councils should have technology grants:

> I think that funding agencies should realise that technology underpins science. I think that realisation is just starting now, but in earlier times I don't think that the Research Councils in any way realised the significance of technology in the same way as industry has always done. (...) And there is still a tendency in the Research Councils to fund technology and capital equipment, but to require the researchers to make do. That is all very well, but what is the point of wasting the time of PhD students or post-docs if they just are going to be uncompetitive and beaten by those based in other countries where research may be funded properly? (interview Bowles, 2006)

Currently, a working solution is found. The facility both provides training courses and performs contract work. In addition, a certain percentage of the grant money of research projects is reserved for the facility. This mixture enables the facility to 'get along'.

In conclusion, the case of the Technology Facility in York is an example of centralisation around technology. But instead of centralisation around one large technology, centralisation takes place around various smaller technologies. Moreover, the gravitation of technologies is not the leading force behind centralisation. Against the background of scientific developments and competition between public and private parties, the Technology Facility was actively constructed by a person with innovative ideas on how technological development should be incorporated in academic research practices that later materialised in new socio-technical arrangements. More specifically, this case shows the creation of a new division of labour between technicians and scientists. Instead of a classic form of labour division, scientific craftsmanship further evolves when scientists perform their own research with the help of technicians in the Technology Facility.

Networking: the integration of information technologies

The third organisational movement does not concentrate research in biology but goes in the opposite direction. In interaction with the increasing importance of information and communication technologies within biology, large-scale research efforts have acquired a networked character. Instead of research being performed in one single location, the places of research are dispersed while connected through information infrastructures. These networked forms of research emerged together with the conceptualisation of biology as information and the integration of information technologies in the biosciences. Nowadays, information and communication technologies are crucial in biology, which is exemplified by the emergence of bioinformatics as a new subdiscipline. Moreover, the increasing importance of data-networks transforms the character of research in biology profoundly. This section explores networking, by investigating the integration of information technologies in biology.

The 20th century can be characterised as the century that elaborated the idea of biology as information (Beaulieu, 2004; Keller, 1995; Kay, 2000). This is explicated by the director of the largest sequencing center in the Human Genome Project, Eric Lander, who presents three views on biology: biology as organisms, biology as molecules and biology as information (Lander, 2003a;

Lander, 2003b).[42] When going back to Mendel, a focus on the function of *organisms* led to attention to heredity, which was later located at the *molecular* level. In turn, the study of 'the world of molecular machines' brought DNA, which formed the beginning of biology as *information*. The purification of the information from the DNA molecule became possible through sequencing:

> DNA is composed of an ordered series of four chemical structures called nucleotide "bases": andenine, thymine, cytosine, and guanine, which are abbreviated A, T, C and G. These bases are lined up one after another along the length of a DNA strand. The sequence of these bases acts as a code that can be deciphered to reveal our genetic instructions. (NIH, 2003: 4)

The metaphor of DNA as information is very powerful (Fox-Keller, 1995; Kay, 2000). DNA as 'the book of life' is often used to explain and sell genomics research to a wider public. For instance, Lander speaks of a storybook: "The genome is a storybook that's been edited for a couple of billion years, and you could take it to bed, like A thousand and One Arabian Nights, and read a different story, in the genome, every night" (Lander, 2001). In addition, DNA is presented as the instruction book for making the proteins that make-up our body: 'the blueprint to build our body' (NIH, 2003: 3).

The integration of information technologies in the biosciences is closely interwoven with this so-called 'informational turn' (Beaulieu, 2004; Groenewegen & Wouters, 2004; Hine, 2006). The conceptualisation of biology as information developed in interaction with the integration of information technologies in biology and the specialisation in bioinformatics. First, taxonomists started to use new information technologies for repetitious tasks and the management and storage of the increasing amount of data in databases. In the 1970s and 1980s computing and computer modelling were introduced in the biological sciences, leading to the emergence of bioinformatics:

> [B]ioinformatics can simply be defined as the application (or integration) of computer science to molecular biology (…) in general, bioinformatics has specialized in three areas of biotechnology research: sequence and structure analysis (genomics, proteomics), data-management (large-scale sequence repositories such as GenBank), and the development of integrated systems for "in silico" biology (simulation systems for the testing of drug compounds). (Thacker, 2005: 54)

---

[42] During the HGP, Lander was director of The Whitehead Institute/MIT Center for Genome Research. Retrieved February 10, 2007 from http://www.wi.mit.edu

Bioinformaticians are a special breed of scientists that have one foot in the world of information science and the other in the world of biology. Although currently educational programmes teach the two disciplines in an integrated way, the first scholars in bioinformatics had an educational background in one of the two disciplines and learned the other along the way.[43]

The integration of information technologies in biology co-evolved with the creation of networked research arrangements (Kohler, 1994; Gaudillière & Rheinberger, 2004; Stemerding & Hilgartner, 1998; Sulston & Ferry, 2002). Ideas on DNA as information are fundamental to sequencing methods that developed within research networks around fly, worm and human. As the scale of the task of sequencing complex organisms as the human being could only be accomplished through a critical research mass, different scientists and groups increasingly started to interact, connecting different national and international research sites. Social forms of coordination were combined with standardisation through technologies. For instance, researchers first used different mapping techniques which did not allow the merging of research results from different laboratories; the construction of the so-called 'sequenced-tagged-site' (STS) made the combination of different mapping techniques possible: "[STSs] offered a way to join data from "any of a variety of physical mapping techniques," reporting it in a 'common language'" (Stemerding & Hilgartner, 1998: 61).

The creation of data-networks has been used as means of coordination between dispersed research labs and across time (Groenewegen & Wouters, 2004; Hine, 2006; interview Siebes, 2004; interview Vriend, 2004). The enormous amounts of data resulting from sequencing and similar analytical research projects are stored and integrated through (online) databases: "in order to benefit in full from the possibility to combine knowledge on a larger scale, knowledge repositories and places of knowledge creation need to be combined (Groenewegen & Wouters, 2004: 167). This materialised in the construction of online genetic databases, as 'GenBank' which is managed by the NCBI at the NIH and 'Ensembl', a co-operation between the European Molecular Biology

---

[43] To illustrate, Professor Gert Vriend, director of the Center for Molecular Informatics at Nijmegen University which forms the basis for the Netherlands Bioinformatics Center (NBIC), has a background in biochemistry but early computer lessons in the 1970s wakened his interest in ICT (interview Vriend, 2004).[43] After doing a PhD in virology, he left for a post-doc to the United States to develop software, and amongst others worked as bioinformatician at the EMBL before landing in Nijmegen. In contrast, Professor Arnold Siebes from the Institute of Information and Computing Sciences at Utrecht University in the Netherlands has a background in mathematics, specialised in datamining and only relatively recently switched to the field of biology (interview Siebes, 2004).

Laboratorium, the European Bioinformatics Institute and the Sanger Institute.[44] However, building connections between different data and keeping data up-to-date requires lots of work on standardisation. Some bioinformaticians are working every single day to keep information connected, allowing other scientists to use the data in their research. However, the biggest challenge for bioinformatics is not the storage of data or making data accessible, but turning data into knowledge.

Currently, various approaches try to turn biological data into knowledge (Breit, 2006; SARA, 2006; interview Siebes, 2004; Siebes, 2001; Vriend, 2000; interview Vriend, 2004). The most important way of making data meaningful is called 'datamining' which consists of data comparison in order to recognise patterns and the integration of information to construct models. For instance, (parts of) the genomes of different organisms are compared in order to find similarities that can tell something about evolution, or relationships between genotype and phenotype can be discovered by looking for differences between the genome of two individuals, the so-called Single Nucleotide Poymorphisms (SNPs). The ultimate goal is to get insight into the structure of parts and processes in the cells, so-called pathways. Part of this new form of life sciences is the simulation and visualisation of information and processes in the cell. For instance, in SARAgene – "a 3-D visualization and datamining tool for genomics related data" (SARA, 2006) – scientists are able to see the relation between different sets of data in 3D, which can lead to new insights. Another example is the virtual laboratory: looking like a control room with a large high-definition, tiled screen on which data can be displayed and reworked. These virtual places can be seen as research nodes within the worldwide networks of data.

The increasing importance of data is transforming the character of research in biology. Historian of bioinformatics professor Timothy Lenoir even argues that the informational turn radically transforms the future of biology:

> Many molecular biologists who welcomed the Human Genome Initiative undoubtedly believed that when the genome was sequenced, everyone would return to the lab to conduct their experiments in a business-as-usual fashion, empowered with a richer set of fundamental data. The developments in automation, the resulting explosion of data, and the introduction of tools of information science to master this data have changed the playing field forever. There may be no 'lab' to return to. In its place is a workstation hooked to a massive parallel computer producing simulations by drawing on the data

---

[44] See respectively http://www.ncbi.nlm.nih.gov/Genbank/index.html and http://www.ebi-.ac.uk/ensembl, retrieved from the web February 11, 2007.

streams of the major databanks and carrying out 'experiments' in silico rather than in vitro. (Lenoir, 1998b cited in Hine, 2006: 270)

Although relocating the laboratory to the dustbin of history seems a little premature, Lenoir describes a move from 'wet' material research in the laboratory to 'dry' information-based research that is recognised more broadly (interview Siebes 2004; interview Vriend, 2004; Chong & Ray, 2002; Fox-Keller, 2005; Stevens, 2004). Or as Lander reformulates this shift: "the browser becomes the new microscope" (Lander, 2004). Studying biological data instead of the actual organisms brings a new research approach into biology, turning biology into a hypothesis driven science. When having a question, scientists can now first try to find the answer by using databases instead of going directly into the laboratory to do an experiment.[45]

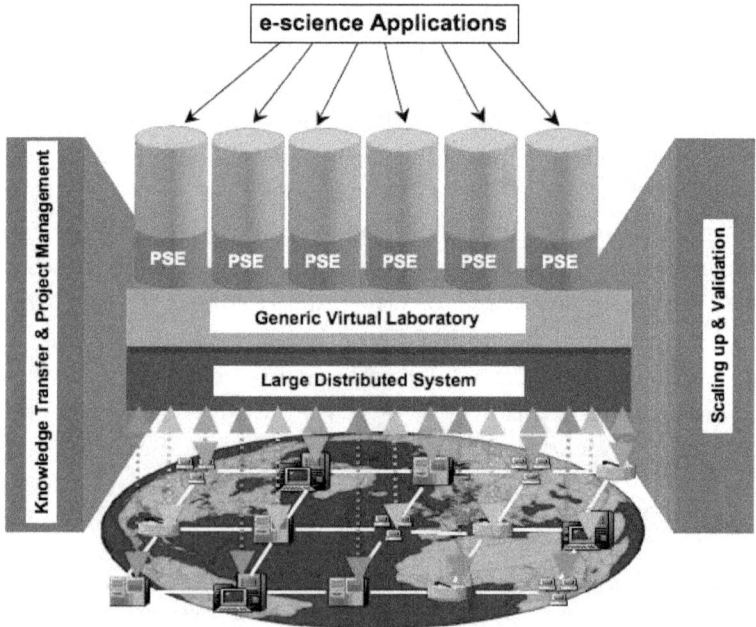

The structure of a virtual laboratory[46]

---

[45] Nevertheless, it is often argued that the laboratory will always be important in biology research, as you still have to test in experiments: "Only through the microscope you can see if you were right" (interview Siebes, 2004).
[46] Image courtesy of the Virtual Laboratory e-Science, http://www.vl-e.nl

In sum, the emergence of research networks in molecular biology co-evolved with the integration of information technologies in biology. Thereby the re-conceptualisation of biology as information has been crucial. Researchers created a new perspective on biology, with important implications for research practices. Not only bioinformaticians have become important actors in biology, laboratory biology also acquired new organisational forms. The construction of sequencing projects involved standardisation processes and the creation of data-networks and this turned molecular biology from a non-collaborative into a collaborative science. Different research sites work together to a common objective, which is enabled by the creation of socio-technical networks making sure that the diverse research results can be integrated in the end. However, the large-scale sequencing projects are not only an example of the networked character of biology research, but the huge amounts of information that they produce also further stimulates network formation in biology and shifts re-search practices towards dry, computer-based research.

Conclusion

The organisational movements around technology in biology only partly re-semble the organisational configurations in big physics. Biological instruments are relatively small and simple when compared to such typical big science instruments as particle accelerators and space shuttles. Consequently, biological instruments do not have the same centralising force as in particle physics or space research where research concentrates around a large instrument as focal point. However, when looking from within the biology field, instruments have certainly become bigger, more complex and more expensive, and this has influenced the organisation of research. These higher levels of aggregation around larger instruments and centralisation of complex and expensive instru-ments resemble centralisation in physics, even though on a smaller scale. In contrast, the networked character that developed in interaction with the integra-tion of information technologies sets big biology apart.

The history of collaboration in biology and the role that technology plays in the formation of these collaborations also underscore that technology is not the only incentive for collaboration. The character of research material, the research approach and political movements towards scientific collaboration shape big biology, too. When looking into specific interactions between tech-nology and organisational transformations in biology, it becomes clear that technologies are part of a dynamic process in which research arrangements are actively constructed by people, organisations and policies. The establishment of new socio-technical arrangements demonstrates how these configurations are continuously reshaped and create new professional roles as well. This is why my analysis resonates with critiques of the technological deterministic character of big science literature (Capshew & Rader, 1992; Westfall, 2003). Still, life scientists often do experience technologies as an important source of transformation, which is only logical when taking into account the pace and scope of the technological developments in biology. The individual scientist is regularly confronted with new, already more or less stabilised technologies that they have to incorporate into their research in order to keep up-to-date.

The societal context of big biology

Because the internal dynamics of science significantly contributes to shaping research arrangements, while external factors have much influence on big science complexes, these arrangements and complexes can only be understood in the light of their social setting. While the former chapter already showed how big physics emerged before and during World War II and was conceptualised in the 1960s, large-scale research arrangements in biology only recently became prominent in quite a different societal context. How does this different societal context influence the specific form of big biology? By reviewing the societal context in which big biology was shaped, I will further specify the differences between large-scale research projects in physics and biology before formulating a final empirical characterisation of big biology.

There is a clear connection between big physics and the emergence of big biology at the end of the twentieth century and the beginning of our new century. In the after-war period large-scale research in physics had to reposition itself, which amongst other things led to the incorporation of biology research. As investments in physics became controversial, first because of the devastating power of the Atomic Bomb and later because of the debate on Nuclear Energy, the national laboratories in the United States started to incorporate research into life in their strategy, because this brought less controversial societal benefits and also fitted with the rising attention for environmental pollution. Thus the large-scale research organisations for physics gave a new impetus to biology research, which became particularly apparent in radiobiology and environmental research (Bocking, 1997; Creager & Santesmases, 2006; Schloegel and Rader, 2005). Moreover, in the context of their interest in the effects of radiation on the human genetic make-up, it was the Department of Energy that first envisioned large-scale research collaboration in molecular biology. This department used its experience with organising large-scale physics research in shaping the Human Genome Project, which opened the way for other large-scale research projects in the context of genomics and post-genomics research. Consequently, at the end of the twentieth century, large-scale research collaboration in biology became more prominent and was staged as big science.

In the meantime, the modern society that gave rise to big physics has been replaced by our contemporary society, which is characterised by various labels. Characterisations such as knowledge society or economy, globalisation and network society most clearly relate to the development of big biology. The transformation from an industrial to a post-industrial society has spawned our knowledge society, one in which science and technology have replaced industrialisation as foundational underpinning of society at large (Bell, 1967; Drucker, 1969; De Wilde, 2001). Obviously, knowledge and its institutions are increasingly important in the shaping of society and everyday life. As biology predominantly creates knowledge for the civilian domain, it is not surprising that biology grows big in the context of this knowledge intensive society. In contrast, big physics and space research largely developed in a context of warfare and rivalry between superpowers (Lenoir & Hays, 2000; Creager & Santesmases, 2006).

Furthermore, our knowledge society is characterised by a transition from an industrial economy towards a knowledge economy. This is clearly reflected in developments in biology research, as knowledge about life has become a commodity (Rose, 2001; Rose, 2006; Sunder Rajan, 2006; Thackray, 1998; Waldby & Mitchell, 2006; Yoxen, 1984). In addition, the international and networked character of big biology resonates with such more recent characterizations as globalization or network society (Castells, 1996; Van Dijk, 1991; Held, 1999; Thacker, 2005). In interaction with the expansion and integration of socio-technical infrastructures, worldwide exchange has increased enormously, resulting in the so-called 'global village'. With a special focus on information technologies in connecting places around the world, Castells speaks in this context of the emergence of a network society in which social structures increasingly have a networked structure – in this case visible in biology research.

The emphasis on processes of innovation and the (industrial) application of science that is central in our knowledge economy also becomes apparent when focussing on the relation between science and society during the period in which biology grows big (Jasanoff, 2005; Remington, 1988; Rip, 1998). After the social contract for science of the 1960s, the 1980s put science and technology forward as a means towards economic growth. This gave rise to so-called 'strategic science' which is defined as "basic research carried out with the expectation that it will produce a broad base of knowledge likely to form the background to the solution of recognised current or future practical problems"

(Irvine & Martin, 1984: 4). Moreover, there was increasing attention to innovation processes and innovation policies. As a result, the protected space for science that was created by the 'social contract between science and society' has now been opened up by pressures for the relevance of scientific research and linkages with the world 'outside' science. Consequently, authors now speak of the 'new social contract' in which science needs to be staged in a context of application.

This emphasis on innovation in science policy sets biology apart from more classical forms of big science, for the requirement of application makes expansion of biology rather complicated. Although the life sciences are a source of medical, agricultural and environmental innovation while the growth of biology seems only natural within the contemporary context of innovation, this particular context also comes with specific challenges. Life scientists experience that the current social expectations of science make the building of large-scale research networks in biology a different game (interview Andeweg, 2005; interview Pierrot, 2007; interview Sibuet, 2006; interview Sinclair, 2006; interview Van Driel, 2005; interview Westerhoff, 2005). Within biology scientists are always expected to know future (industrial) applications of their research, while at the same time they observe that big physics and space research receive research funding without the societal usefulness of their projects being clear from the outset. As one of my interviewees commented: "Even when these space shuttles explode, they still receive huge amounts of money." The solid organisational machinery of these classic forms of big science, built in times of the old social contract, and their extensive lobbying experience can be seen as reasons why big physics and aerospace research manage to get such a big piece of the pie. This poses the challenge to biology to build its bigness within an environment that values application over more basic forms of science; but biology also has an organisational disadvantage vis-à-vis the older forms of big science.

Finally, big biology research has been an important factor in the broader movement towards the embedding of science in society (Nowotny et al., 2001). Biology's potential for use legitimises large-scale biology research, but also makes it more complicated as it touches on important issues, such as health, food and the definition of life. This implies that biological research and its applications are often contested, as controversies about genetically modified food or stem-cell research show. Although big physics has given rise to some serious controversies as well, in the context of big biology these issues first became addressed in an integrated way. The Human Genome Project was linked to an extensive programme on the ethical, legal and social implications of genetic research, which set the stage for many more research programs into the interaction between biology and society (Vullings & Planque, 2002). As a result, the embedding of science in society has become more important, in particular within the context of biological research.

Biology as a networked form of big science

Recent transformations in the life sciences have turned biology into big science. Although biology has a substantial history of scientific collaboration, it has traditionally not been presented as part of the big science family as laboratory biology always was small-scale. However, the recent increase of collaboration in laboratory biology is presented as big biology. This implies a distinction can be made between traditional forms of collaboration around the assembling of biological material and, later, small-scale analytical and experimental laboratory research on the one hand, and recent techno-scientific collaborations that explore the molecular basis of life on the other. In line with Pickstone's ideas of different ways of knowing (Pickstone, 2000), these different ways of doing biology research build on each other and also exist next to each other.

When comparing big biology to traditional big physics, it becomes apparent that collaboration in biology does not exactly resemble big physics: biology is a different form of big science. To be more precise, I have argued that big biology stands apart through its networked structure and the integration of information technologies. In addition, I have shown how big biology has emerged in interaction with contemporary society. Nowadays, the generation of knowledge is fundamental to societal and economic developments, which has become established in the 'new social contract' between science and society. Finally, the emphasis on ethical, legal and social implications makes big biology firmly embedded in society. This characterization of big biology sharply contrasts with more classical forms of big science of which traditional big physics is exemplary (see Table 2). As a result, big biology can be put forward as a contemporary, networked form of big science.

Centralised big science	Networked big science
centralised organisation	decentralised organisation
concentration of research around large and expensive instruments	networked collaboration underpinned by transport and information technologies
emerged in beginning of 20th century in context of physics	emerged in 17th century in context of scientific explorations
dominant in 20th century in physics and space research	prevailing nowadays in (molecular) biology in which ICTs are integrated
part of modern society characterised by belief in progress, growth, rationalisation, industrialisation and bureaucratisation	part of contemporary society characterised as post-industrial, knowledge society/economy, globalisation and network society
science as source of power	science for economic growth
industrialisation of science	embedding of science in society
social contract between science and society: protected space for (basic) science with continuous government support	new social contract between science and society: emphasis on application, innovation and societal implications of science

Table 2: Centralised big science versus networked big science

While in this conceptual part of the thesis I have characterised big biology as a different form of big science compared to big physics, in its empirical part I will refine this characterisation of scientific collaboration in biology. 'Scientific collaboration' and 'biology' already intertwine in a number of detailed studies. On the qualitative side one finds not only historical case studies of scientific collaboration in the life sciences, but also studies of the possibility and impossibility of collaboration within today's laboratories (Hackett, 2005; Kleinman, 2003; Knorr-Cetina, 1999; Fujimura, 1996; Rheinberger, 1997). These studies have to be placed in the tradition of laboratory studies (Latour & Woolgar, 1986; Latour, 1987). When transgressing the walls of the laboratory, accounts of scientific collaboration in biology are mostly written in the context of the Human Genome Project (Balmer, 1996a; Balmer, 1996b; Gaudillière & Rheinberger, 2004; Glasner, 1996; Kevles & Hood, 1992; Sloan, 2000). In turn, I will broaden the empirical view on contemporary scientific collaboration in biology beyond the Genome Project, by investigating other contemporary research collaborations in biology.

The following three chapters will present three different styles of collaboration in biology, combining a specific way of knowing with a particular organisation of research.[47] First, the Census of Marine Life pursues classification and cataloguing of life in the oceans and can therefore be characterised as contemporary natural history collaboration. Secondly, the Silicon Cell project, which aspires to build a cell model in a computer, is an example of the construction of collaboration within laboratory biology. Where the Human Genome Project was primarily an analytical project, the Silicon Cell is dedicated to the building of models on the basis of analysis and experimentation, and as such seeking to become even bigger than the Genome Project. Finally, the VIRGO consortium, an academic-industrial collaboration, is committed to application. The project investigates host-virus interaction in order to develop new diagnostics and medicines for infections such as influenza.

Because these various styles of collaboration imply different processes of integration, I approach each case study from a different perspective. With respect to the Census I take a historical perspective, looking at continuity and change in natural history collaborations, emphasising the development and integration of different forms of collaboration over time. In case of the Silicon Cell I adopt a sociological perspective, asking how scientific collaboration is built. With the integration of research units into a networked form of collaboration, I explore the emergence of collaboration in analytical and experimental ways of knowing in biology. Finally, I capitalise on theories of innovation to investigate the integration of academia, industry and government in the VIRGO consortium.

---

[47] I will further elaborate on the different styles of collaboration in Chapter 6.

PART 2

Life sciences live

CHAPTER 3

Seeing life in the oceans
New natural history

Venter as the new Charles Darwin[48]

While the discovery of space is well under way and almost every piece of land in the world is discovered and mapped, not much is known about the world's oceans that cover about 71% of the earth's surface.[49] Next to life in the depth of the oceans, especially invisible life forms like micro-organisms are still a big mystery. Therefore, after finishing the map of the human genome, Craig

---

[48] This picture of Craig Venter has been taken onboard of the Sorcerer. Photographer unknown.
[49] According to the Encyclopaedia Britannica online. Retrieved, August 23, 2007 from http://www.search.eb.com

Venter's new adventure entailed the discovery of micro-organisms in the oceans. In 2003 he embarked on his ship the *Sorcerer II* and set-off to the Sargasso Sea taking water samples to sequence all its micro-organisms. [50] As this small expedition already revealed an incredible degree of microbial diversity, Venter announced the 'Sorcerer II Global Ocean Sampling Expedition'. This explains why he is pictured as the new Charles Darwin collecting the DNA of everything on the planet (Shreeve, 2004a). But the expedition of Venter is only very small when compared to the 'Census of Marine Life' (CoML). This large-scale research project does not only want to reveal micro-organisms, but tries to catalogue all the animals in the world's oceans, including life in the deep-sea. This large-scale international project involves over 2000 scientists from more than eighty countries. The objective of the ten-year project that officially kicked-off in 2000 is "to assess and explain the diversity, distribution, and abundance of marine life in the oceans – past, present, and future".[51]

Expeditions that explore life in the oceans are part of a natural history tradition in which collaboration is necessary for gathering research materials. While the Human Genome Project is often presented as the first large-scale research project in the life sciences, scientific collaboration, as argued in the previous chapter, is hardly new to biology. It is found already in the alliance between science and exploration that set out to map the world and collect and describe its diverse forms of life. Studies of scientific collaboration pay little attention to these collaborations that collect, identify and catalogue life on earth. If ecology is gradually becoming an important subject of studies into big science (Bocking, 1997; Kwa, 1987; Parker, 2006; Schloegel & Rader, 2005), research into the world's oceans is not taken into account yet. In contrast, this chapter focuses on large-scale marine biology research as an example of large-scale natural history. More specifically, I study the contemporary Census of Marine Life to explore transformations in natural history collaborations. Does this current collaboration still resemble traditional natural history collaborations, or did the expansion of molecular biology research and recent changes in science and society also transform traditional large-scale biology?

---

[50] See Venter et al., 2004 and press release from JCVI 'More than Six Million New Genes, Thousands of New Protein Families, and Incredible Degree of Microbial Diversity Discovered from First Phase of Sorcerer II Global Ocean Sampling Expedition'. Published online, March 4, 2004. Retrieved August 23, 2007 from http://www.jcvi.org/press/news/news_2007_03_13.php In addition, information can be found on the website of the expedition and the J. Craig Venter Institute. Retrieved August 23, 2007 from http://www.sorcerer2expedition.org and http://www.jcvi.org.

[51] 'About the Census of Marine Life' from the website of CoML. Retrieved, August 23, 2007 from http://www.coml.org/aboutcoml.htm.

A comparison between the expeditions of Darwin, Venter and CoML suggests a similar natural history approach in some respects, but also shows several transformations. Traditionally, natural history took place in the context of the Renaissance, trading nations, empire building and the establishment of scientific societies and national museums (Pickstone, 2000). When Darwin travelled the oceans on the HMS *Beagle*, he primarily studied and collected life at the sea shores and specimens were pictured or stored in so-called cabinets of curiosity and used to build theories about the history of life. In contrast, contemporary explorations take place in a globalising world predominantly shaped by know-ledge and technology (Bijker, 2006; Etzkowitz & Leydesdorff, 2000; Gibbons, et al., 1994; Rip, 2001). Venter now classifies invisible species, not by their appearance but by using genomics technologies while data are stored in an electronic database.[52] In addition, CoML studies all animals in the sea – from microorganisms to whales – thereby covering all areas in the ocean, from the shores to the deep-sea. Scientists make use of old-fashioned sampling methods as well as the newest techniques available and the research results are used to predict the future of life in the oceans. Moreover, the ocean is seen as a source of innovation as the application of research is emphasised.

That the Census of Marine Life can be compared in some respects to more traditional forms of scientific collaboration in natural history allows me to study transformations in this style of collaboration. This chapter, then, provides a historical perspective by investigating issues of continuity and change. According to sociologist of science Arie Rip (2001) the new natural sciences, including biology, are still measuring, mapping and modelling the world – as the natural sciences always did – but now in a more sophisticated way, due to developments in information and communication technologies. In the context of the Census of Marine Life, I will explore what this 'sophistication' actually entails. What kind of transformations take place exactly? I will argue that the integration of information technologies is only part of the transformations that are taking place in natural history types of collaboration. Although continuities with the past can be observed, contemporary developments in big biology that are outlined in Chapter 2 have also transformed traditional natural history collaborations. Next to increasing scales, new socio-technical infrastructures are built and research becomes more strongly embedded in society. My analysis of the Census shows how a contemporary natural history project is indeed still

---

[52] While Venter became known as the private party in the public-private race for the sequence of the Human Genome, he has now established a not-for-profit research institute. The data from the Sorcerer expedition are stored in public databases and the project is connected to the larger CoML.

measuring, mapping and modelling the world, but this work has changed substantively in interaction with recent scientific, technological and societal developments.

In order to discuss issues of continuity and change in the Census of Marine Life, this chapter will first give an overview of the history of ocean research, showing how scientific, technological and societal factors have interacted towards the development of large-scale collaboration that is required to reveal life in the oceans. Next I will look at current research, addressing the history, structure and achievements of the Census of Marine Life. This implies I will examine transformations in natural history by identifying and discussing loci of change. These are found in globalisation, building a new information infrastructure, technology development, reinventing taxonomy, applying marine biology and showing the public. Finally, the conclusion will reflect on the characteristics of new natural history. [53]

Exploring the oceans

The Census of Marine Life is one of the two large-scale international projects in present-day marine biology. When I first accessed the Census website in 2004, the collaboration entailed three hundred scientists from fifty-three countries. Later this turned into one thousand scientists from seventy countries and participating scientists are working to involve as much scientists and countries as possible. While I am writing this chapter more than two thousand scientists from eighty nations are engaged:

> The Census of Marine Life is a growing global network of researchers in more than 80 nations engaged in a ten-year initiative to assess and explain the diversity, distribution, and abundance of marine life in the oceans – past, present, and future.[54]

Next to the number of scientists and countries involved, the project's finances of more than $ 1 billion, the ten-year duration and the research goals of the

---

[53] My analysis of the Census of Marine Life is based on primary and secondary literature on the collaboration, consisting of websites, policy documents, meeting minutes, scientific articles and media coverage of CoML. In addition, I have performed in-depth interviews with a number of participants with various positions within CoML (for a complete overview of interviews, see Appendix A). Moreover, I have observed meetings concerning ocean and biodiversity research (for an overview of attended meetings, see Appendix B).

[54] From the webportal of CoML. Retrieved August 14, 2007 from http://www.coml.org.

project give an idea of its scale and scope.[55] CoML not only aims to catalogue all current life forms in the oceans in a large database, but also wants to get an overview of what has lived in the oceans in the past and what will live in the oceans in the future.

Research into the oceans and their living creatures is big science avant la lettre (Ballard & Hivly, 2000; Dunbar, 1971; McIntyre, 2005; Menard, 1969). "Marine studies in general have a very early history in collaboration. It is essentially big science" says Michael Sinclair, director of Canada's largest centre for ocean research (interview Sinclair, 2006).[56] The dispersed nature of the research material is traditionally an important reason to collaborate. Marine science has an international character as oceans are spanning the entire globe.[57] In addition, collaboration between disciplines already has a long tradition within marine science. As the director of the deep-sea research department of the French research institute for exploitation of the sea, Myriam Sibuet states: "The multi-disciplinary character of problems asks for collaboration between for example biologists, physicists and chemists" (interview Sibuet, 2006).[58] Also the technology that is needed to explore the oceans is presented as main reason to collaborate: "The instruments of marine science can be compared to the large and expensive instruments that are used in physics and astronomy, like huge telescopes and cyclotrons" tells the director of the Royal Netherlands Institute of Sea Research Carlo Heip (interview Heip, 2006).[59] For instance, a research vessel costs about $60 million nowadays.[60] However, ocean research has not

---

[55] Although the round number of $ 1 billion was put forward by the scientists at the beginning of the programme, towards the end of the programme the Census community estimates that the actual expenditure will be between $ 600-650 million (personal communication with Jesse Ausubel, January 25, 2009).

[56] Prof Dr Michael Sinclair is director of the Bedford Institute of Oceanography (BIO) in Dartmouth Canada. He is a member of the Scientific Steering Committee of CoML and participates in the Gulf of Maine Census, one of the pilot projects of CoML.

[57] To illustrate, the International Council for the Exploration of the sea originated more than 100 years ago, when the Swedish Otto Petterson recognised that Sweden and Norway were not able to address questions of climate change or fish population alone and engaged more European countries to explore these issues (interview Sinclair, 2006; Rozwadowski, 2002).

[58] Prof Dr Myriam Sibuet is director of the 'Département Environnement Profond' of the 'Institut français de recherche pour l'exploitation de la mer' (Ifremer) in Brest, France. She is member of the Scientific Steering Committee of CoML and programme leader of CoMarge.

[59] Prof. dr. Carlo Heip is director of the Centre for Estuarine and Marine Ecology from the The Netherlands Institute of Ecology (NIOO-KNAW) and general director of the Royal Netherlands Institute of Sea Research (NIOZ). He is a member of the Scientific Steering Committee of CoML and founder of the Euro-COML committee.

[60] This number is based on the recent acquisition of the G.O. Sars, a Norwegian vessel that is part of CoML (interview Heip, 2005).

always been big. This section will show how interaction between scientific, technological and societal factors transformed ocean research into big science. After starting with short history of ocean research, I will especially focus on developments in marine biology that paved the way for the Census of Marine Life.

The great scientific voyages of the 18th and 19th centuries – including Charles Darwin's famous journey aboard the HMS *Beagle* – only explored species near the surface of the ocean, because they had neither access to nor knowledge of the deep oceans. It was only towards the 20th century, that the development of techniques enabled the exploration of the deep and the ocean floor (Kunzig, 2000). At first, scientists began to investigate the depth of oceans. Sound to measure depth was first ventured by the Swiss mathematician Colladon in the Lake of Geneva, using a church bell and an ear trumpet. In 1838 this method was transferred to the ocean using explosions. In 1853 this developed into the so-called 'soundingline of Brooke' which was employed by the US Navy Depot of Charts and Instruments to map the North Atlantic Ocean. Thereby they discovered the "telegraphic plateau" between Newfoundland and Ireland, which would be used by the Atlantic Telegraphy Company for the first trans-oceanic cable in 1858. In the 1870s the crew of the converted warship *Challenger* – known as the mothership of oceanography – discovered the Mid-Atlantic Ridge. As a result, scientists slowly began to realise that the oceanfloor had similar characteristics as the earth's surface, and in 1904 the newly established International Hydrographic Bureau published the first bathymetric standardised chart of the world ocean, based on 18400 soundings.

In the 20th century, ocean research gradually professionalised. Next to telegraphy, shipping traffic and the Titanic disaster were important incentives to develop new technologies to survey the oceans (Kunzig, 2000; Oreskes, 2003; Theberge, 2006). In addition, the World Wars and the following Cold War brought submarine warfare. In these contexts various new instruments were constructed and others came into widespread use. Technicians would design for instance sophisticated navigation instruments and deep-sea cameras. Moreover, ocean research became institutionalised. To illustrate, the American Woods Hole Oceanographic Institution (WHOI) which today is the world's largest private, non-profit ocean research organization, grew substantially during the war period. This institute, founded in 1930 with funds from the Rockefeller Foundation, played an important role in the development of oceanography.[61] The history of the research vessels developed by the institute gives a nice

---

[61] 'History of the WHOI research vessels'. Retrieved August 23, 2007 from http://www.whoi.edu

overview of the evolution of research vessels from traditional small sail and steamer ships to modern research vessels. The difference between their first and most recent research vessel can be seen in the pictures.

The first research vessel *Atlantis* (1931-1964) and the most recent *Atlantis* (1997)[62]

As marine biology developed in close interaction with these more general explorations of the oceans, actual observations of life in the deep-sea are of fairly recent date. For a long time it was thought that life could only be found at the shores and the ocean surface as the absence of light, low temperatures and the density of water in the deep ocean would not allow any life (Kunzig, 2000). However, these ideas slowly changed with the development of technologies that made life in the deep ocean visible. In the 19th century the Irishman Forbes developed and professionalized the art of dredging in order to explore life in the deep. Later, scientists from the Whoods Whole Institute combined a dredge and trawl into a so-called 'epibenthic sled', used for bringing the diversity of life in the deep oceans to the surface. In the course of the 20th century, scientists have increasingly gained access to the deep ocean, facilitating direct observation of life in the deep-sea (Ballard & Hivly, 2000; Kunzig, 2000: Oreskes, 2003). The development of the 'bathysphere' – a kind of underwater balloon – in the 1930s enabled the first observations of life. Later, in the 1960s, the 'bathyscape' – a kind of underwater zeppelin for two men – descended to the Challenger Deep of the Mariana Trench (at about 11 kilometres the deepest known spot in the oceans). They spend 20 minutes there and saw a fish, which indicated that even in the deepest ocean life is possible. The construction of the submersible

---

[62] Images courtesy of Woods Hole Oceanogrphic Institute.

with robot-arms *Alvin* in 1964 gave researchers even better access to the depths of the ocean.

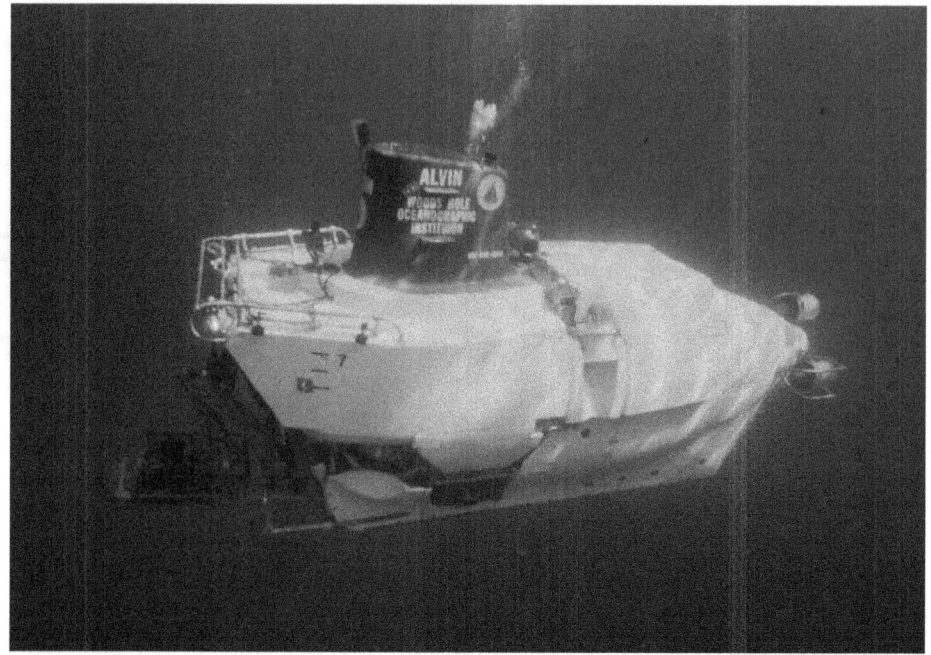

Alvin[63]

In sum, investigations into the oceans have developed through interaction between scientific curiosity, societal exploitation of the sea and technological developments. From the 1960s onwards marine science increasingly became an academic endeavour and the 1970s were even pronounced to be the decade of ocean research. As a follow-up of the International Geophysical Year, the 1970s were arranged to be the International Decade of Ocean Exploration (1971-1980). This scientific decade explicitly aimed to scale-up ocean research to an international level (Menard, 1969). Although it gave an impetus to ocean research and marine biology, attention and funds for this branch of science is still very small compared to, for instance, research into space (Snelgrove &

---

[63] Image courtesy of Whoods Hole Oceanogrphic Institute. Alvin has its own website, which includes a video of a disassembly of Alvin. Retrieved August 23, 2007, from http://www.whoi.-edu/marops/vehicles/alvin/index.html

Grassle, 1997; Grassle & Stocks, 1999; interview Heip, 2006; interview Sinclair, 2006). Nevertheless, today's debates on climate change and biodiversity have granted more prominence to ocean research, and they are an important inspiration for the contemporary Census of Marine Life.

Towards the Census of Marine Life

Building on these general developments in ocean research and marine biology, the Census of Marine Life was put together at the end of the 20th century. This section will analyse the construction and goals of this contemporary natural history collaboration in preparation of analysing transformation in the natural history style of collaboration in biology. After looking into the origin of the collaboration, I will show how this global collaboration is coordinated and what it has achieved so far.

The origin of the collaboration

The Census of Marine Life has a story of origin, or a birth-drama as Knorr-Cetina calls it (1999). The story is about two men who meet, discuss issues of biodiversity and come up with the idea of counting all the fishes in the oceans (Malakoff, 2000). Michael Sinclair, taking part in one of the pilot projects of the Census and member of the Scientific Steering Committee, recalls he project's birth:

> I think the origin of the project is interesting. I personally have only been aware of the origin of two big science projects and in both cases they have really become initiated by individuals who had a vision. In this case it was Jesse Ausubel, who is not a marine ecologist, but rather a programme officer with the Alfred P. Sloan Foundation and a Professor in human ecology at the Rockefeller University in New York. And the second person is Fred Grassle and he was a professor in benthic ecology at Rutgers University. And I think in summertime they both used to spend time in cottages or homes that were close by and they became friends. And Fred's feeling was – and he is very passionate about it – that there was not enough focus on biodiversity and these issues. And in discussing this with Jesse, Jesse said: well let's do something about it. And he had possibilities through his position being in the Sloan Foundation, which is quite unusual for a foundation as they don't solicit projects but they decide themselves if they are going to do something big and then they go and ask. So that's how it started: two men meeting over a beer or something in the summertime. And the right people. (interview Sinclair, 2006)

Fred Grassle and Jesse Ausubel are not only seen as the project's initiators but they still perform major roles. Fred Grassle played a vital role in the development of the project's database and was chair of the Science Steering Committee of CoML until 2008 but still remains very active within the prgramme. In addition, Ausubel plays an important role in the crafting of the CoML initiative. He wrote the initial plans for the collaboration and is responsible for the project's public attention.

While Grassle pursues an academic career, Ausubel has a background in science policy. Fred Grassle's interest in marine biology was first triggered by a biology teacher studying marine invertebrates and as undergraduate he was invited to study the mysteries of life at the sea bottom. Later he spent an important part of his career at the Woods Hole Institute specialising in benthic ecology. In 1989 he founded the Institute of Marine and Coastal Sciences (IMCS) at Rutgers University, where he now is Emiritus Director. In 2005 Grassle, for his major contribution to ocean science, was awarded the 'Grand Prix des Sciences de la Mer Albert 1er de Monaco'. In contrast, Jesse Ausubel is primarily interested in science policy, technology development and public communication. In 1977 he started working at the National Academies of Science being involved in atmospheric and climate studies. He was one of the main organizers of the first United Nations World Climate conference which dealt with global warming issues. In 1989 he became director of the Carnegie Commission on Science, Technology and Government and he currently works as a research associate and director of the Program for the Human Environment at Rockefeller University. Since 1994 he has also been serving as program director at the Alfred P. Sloan Foundation in New York.[64] This foundation was established in 1934 by the president of General Motors, Alfred Sloan, to support research in science, technology and economy. In sum, Grassle and Ausubel combine an interest in biodiversity with scientific and policy expertise and access to the Sloan Foundation, which provides an important part of the funding of CoML.

After the two men came up with the idea of counting the ocean's fishes they started to set-up the project that later became known as the Census of Marine Life. When the Board of Trustees of the Sloan Foundation agreed with a two year feasibility study for a Census of the Fishes in December 1996, Ausubel wrote the primary document that sketches the Census initiative: *The Census of the Fishes: Concept Paper* (1997). The purpose of this concept paper was: "To explore the value, timeliness, and feasibility of stimulating, designing, and

---

[64] Retrieved May 10, 2008 from http://www.sloan.org

organizing a period of intense, comprehensive oceanic observation whose purpose would be to assess and explain the global distribution and abundance of marine life" (Ausubel, 1997). The document presents the first ideas on the initiative and outlines the societal relevance of the study of marine life and concludes that:

> Overall, the time appears ripe to consider a much better mapping of the bio-geography of the oceans, including species composition and time history, in short, a Census of the Fishes. Such a Census might yield not only greatly increased public and scientific understanding of the oceans but also information useful for management of fisheries, marine pollution, and other ecological concerns. (idem)

Finally, the concept paper determines further actions: evaluating the possible purposes of a Census of the Fishes, making plans on how to construct the project, estimate the costs and find out "whether the wish and will exist to do it" (idem).

The concept paper formed the basis for discussions in the marine biology community geared to developing the initiative and finding support. Grassle and Ausubel started to spread their ideas by giving talks during various meetings and conferences and by organising special gatherings: "Our emphasis has been to encourage conversation and debate rather than the creation of documents" (Ausubel, 1999a). The process of consulting scientists began in March 1997 with a gathering of about twenty of the world's leading fish scientists in California (Ausubel, 1999a; Ausubel 1999b). During 1997 and 1998 about three hundred experts from about twenty countries and dozens of different universities and research institutions participated in discussions about the project and various international organisations were consulted.[65] According to Annelies Pierrot who participates in the Census as a taxonomist, this participatory approach is crucial to get the commitment of researchers:

> Consultation with scientists is very important. A top-down approach to a large-scientific project is always doomed to fail, because when you tighten up on grants, people stop doing the research. When you take into account what

---

[65] This included the Food and Agricultural Organisation and the Intergovernmental Oceanographic Community of the United Nations, the International Council for the Exploration of the Sea, the Scientific Committee on Oceanic Research, World Wildlife Fund and the European Union.

researchers want, what they also do without money but then on a smaller scale, this guarantees success and continuity. (interview Pierrot, 2007) [66]

To illustrate, Pierrot tells how in the Census scientists have been able to amend the initial concentration on fishes, broadening the project to the Census of Marine Life. This interaction with scientists and other relevant actors gives the initial phase of the project a participatory character, which can also be found in the governance structure of the project.

Coordinating marine biology research

In December 1998, the Sloan Foundation's Board decided to move from assessing the feasibility of a Census to trying to make the Census happen (Ausubel, 1999a; Ausubel 2001; Baker-Masson, 2000; Decker & O'Dor, 2003; Malakoff, 2000). First, the governance structure of the project was shaped. In 1999 the international Scientific Steering Committee (SSC) was formed: the governing body of Census of Marine Life that provides conceptual guidance, determines the scientific goals, and oversees the progress and direction of the program. The committee meets three times a year and members are also expected to attend conferences as Census delegate to gather the different participants and make the project visible. The steering committee is supported by a central secretariat. Later regional coordination nodes were added, including a European branch.[67]

Together with the establishment of the governance structure, a rough planning for a ten-year project has been made, comprising two years of development. Pilot field projects where planned from 2002 to 2004, followed by the main field projects from 2005 to 2007. The analysis and integration of information were envisioned from 2008 to 2010. In 2000 the project officially kicked-off when the first large grants were awarded to eight research groups involving sixty-three institutions and in 2001 the Census of Marine Life was up and running: "We basically have been able to enthuse some key scientists within various countries about the idea. And they started to search for money to support the projects" (interview Heip, 2006). Now, the Census comprises seventeen global projects. First of all, fourteen fieldprojects look into current

---

[66] Dr Annelies Pierrot-Bults of the Zoological Museum of the University of Amsterdam is a member of Mar-Eco, CMarZ and the bar-coding working group.

[67] For an overview see 'national and regional activities and committees'. Retrieved, May 2, 2008 from http://www.comlsecretariat.org/Dev2Go.web?id=247613 For an overview of the European activities: http://www.eurocoml.org/

life in the oceans, varying from the deep-sea to the shores and from Antarctic life to coral reefs. The results are catalogued by one project that is geared to building a database. Finally, two projects map the past and present of life in the oceans: the History of Marine Animal Population and the Future of Marine Animal Populations.

As a result, the Census exists of a patchwork of projects that are held together by the central governance structure. The funding of research is an important reason for the decentralised governance approach. The field projects that are the main expenses of the Census, costing about $ 5-25 million each, are not covered by the Sloan Foundation but by various national and other re-search funds. The foundation only provides so-called 'seed money', covering starting costs, administration costs, the organisation of meetings, representation and outreach. In addition, the various projects still need to find their own resources:

> While Sloan and other private funders can catalyze the Census, most of the support will need to come from government agencies concerned with science, with fisheries, and with environment, as well as organizations such as the World Bank dedicated to capacity building in developing countries as well as with implementation of agreements such as the Convention on Biodiversity. (Ausubel, 2001)

As all these funding sources have their own coordination requirements, the set-up of the field projects differs, which is enabled by the decentralized structure. Moreover, this structure is seen to fit the broad and diverse goals of the project: "Given the nature of the Census and the very broad objections I think the decentralised approach was appropriate" (interview Sinclair, 2006).

Seeing life in the oceans

Although research results can take quite some time within marine biology, the first results of the Census of Marine Life are starting to appear.

> After an expedition it takes about two years to elaborate the results and being able to publish. So you are always two or three years behind. But it starts com-ing now. We had an all-programme meeting in Frankfurt last year and we will have another next year and then you see all kind of interesting results coming-up. When I extrapolate this, I think: yes, this is going to be quite successful. (interview Pierrot, 2006)

On the project's website and in newspaper articles published worldwide it is shown how through the Census of Marine Life the mysterious world of deep-sea life is slowly uncovered. Especially the various pictures of newly discovered species are fascinating to look at. Results from research in the depths of the North Atlantic Ocean can for instance be admired in the book and public exhibition entitled: Deeper than light. [68]

Picture of Haplophryne mollis as shown on the exhibit[69]

Moreover, the project has transformed the lives of the scientists involved. First of all, it enabled the making of connections: "The programme is about the connections (…) The Census is only possible if you are a community and you share the same language and the same world" (interview Sibuet, 2006). Connections made within the Census are geographical – bringing researchers from diverse countries together – as well as epistemological, bringing disciplines together in a multi-disciplinary effort, fostering diverse research questions and approaches. Connections are made on the governance level – in the Science Steering Committee and the regional nodes – and in the various research parts

---

[68] Retrieved May 2, 2008 from http://www.mar-eco.no/exhibition. See also the section *Showing the public.*

[69] Image courtesy of the photographer David Shale.

of the Census. Although the scientists within a project often already know each other, the collaboration develops the contacts:

> You are able to work with the same samples, with the same goals. For example, we work together with a large group on zooplankton and we worked together on the cruise to gather the samples and now we are also going to work together in the lab to analyse the samples. In this way you can sort things out together and discuss strange things you encounter. (...) in this way the relationships become clearer, you have more insight in the connections. (interview Pierrot, 2007)

In other words, within the Census colleagues become collaborators.

In addition, the Census has helped scientists to find funding for their research. The seed-money of the Sloan Foundation is unique within marine biology:

> Although you have to generate money for the research yourself, you get lots of support, which is of course why the Census has become such a success. Because there is quite a lot of money available for support and pilot projects etcetera. (...) Money generates money and the money of the Sloan Foundation helps to get local and national funding agencies on board. (interview Pierrot, 2007)

However, while all the projects have been able to find funding for the research, some projects get large amounts of money whereas other projects are less successful and have a lower-profile because they cannot find sufficient funds. This makes the planning of the Census difficult:

> It is a fact, rather than a critique, that it is all quite ad-hoc. The success of some parts is not planning but pure luck. Someone has to be interested in the subject and must have been able to secure funds (...) So the programme is basically a mix of projects that have got money. Ideally you would organize things better (...) But you cannot expect the Sloan Foundation to put even more money in it. (interview Heip, 2006)

The funding structure enables scientists to perform their research, but at the same time forms an obstacle to consistent planning and continuity.

Next to funding problems, the project is confronted with the impossibility of fulfilling its goals towards the end of the ten-year period. The final phase proves to be difficult, as data of the various projects have to be integrated, which turns out to be extremely difficult. "This is a big problem, in particular along coasts. Think of all these corals, that is so much, it's just incredible. It's simply impossible to integrate. It is not that we do not try, but it just does not work out" (interview Heip, 2006). To support the integration process, the Census has set up a special framework commission. This commission has to construct a format for the integration: "The form of that will be interesting to see (...) I think that what will happen is that each project will be interesting, each fourteen of them. What is undecided is what will the aggregate look like" (interview, Sinclair, 2006). In addition to the general commission, the individual projects are making synthesis plans and meetings are organised to discuss the way in which results can be integrated.[70]

Not just integration poses a problem; also completing an inventory of all animals in the sea by 2010 is not feasible. "How can we do a census of everything, past, present and future and explain it?" (interview Sinclair, 2006). Consequently, the continuity of research is already on the agenda:

> That is one of the issues we are now discussing. What will happen after 2010? How can we make sure that continuity is guaranteed and that the database will be maintained? And who is going to take care of that and how could it be financed? (interview Pierrot, 2007)

The future of the project is difficult to predict. According to Sinclair the database will carry on long-term, but with the individual projects "we have to wait and see" (interview Sinclair, 2005). However, participants in the Census are expected to engage in the construction of further research plans:

> At the start we said that we will stop in 2010. This is reasonable of course, especially towards the funding organization: we will not knock on your door forever. (...) However, there will certainly be a new research proposal to continue the research, because research into space also never stops. Research into the oceans, in comparison to other research, is still quite small and we still have big knowledge gaps we try to fill. Hopefully, we will be able to do that with the help of the Sloan Foundation or another funding source. (interview Heip, 2006)

In conclusion, as its research goals will certainly not be reached in 2010, it is likely that the project will continue to live on in some form.

---

[70] I observed the difficulty of integrating research results only within benthic ecology research during a project meeting on September 22, 2006 at the ICES Annual Science Conference. As ten of the fourteen field projects within CoML are related to benthic ecology, the objective of the meeting was to come-up with 'cross-cutting issues'.

Transforming natural history

With its objective to catalogue life in the oceans, the Census of Marine Life can be defined as a natural history style of collaboration in biology. Although the project resembles the natural history way of knowing that Pickstone describes, the Census research is performed today. Consequently, it also incorporates recent scientific and organisational developments in biology that I outlined in Chapter 2, which transforms the character of natural history collaboration. This section, will explore these transformation, by identifying various loci of change. Subsequently, I will explore increasing scales, transforming research practices, technological developments, the construction of new socio-technical arrangements, the modelling of life, the application of research and public communication.

Becoming global

Although natural history has been a collaborative effort from its start, the Census of Marine Life has unprecedented global ambitions. With the oceans covering the largest part of our planet, space stimulates the expansion of the Census:

> Because of their area, volume, and diversity of life, the world's oceans are the dominant component of the biosphere. Thus, an assessment of life on Earth must in major part be an assessment of life in the world's oceans. (the Census website)[71]

The Census wants to cover all the world's oceans as well as the diverse areas within these oceans. The project divides the ocean into six realms: from the coastal zone to the deep-sea, from the light zone close to the surface, to the dark zone reaching down to more than 5000 metres and from the ice oceans to seamounts, vents and the microsopic ocean. Together, the fourteen field projects of CoML intend to cover as much of the oceans as possible.

---

[71] Retrieved August, 2007 from http://www.iobis.org/about/index_html#history

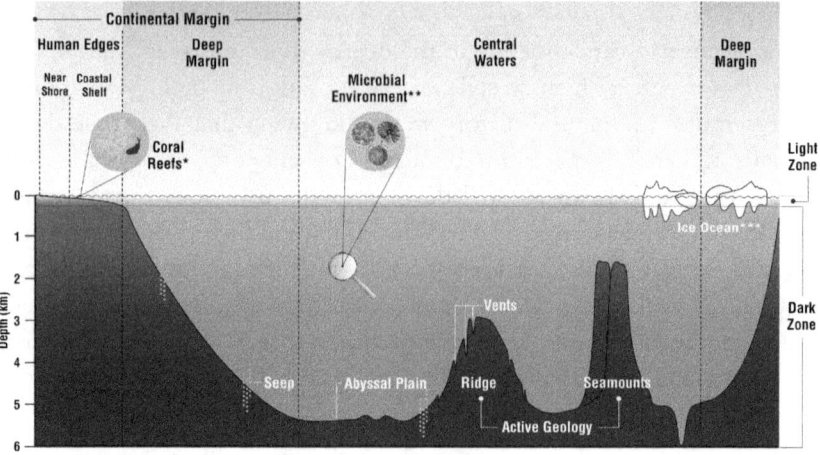

Theoretical Cross Section of the Ocean

* Coral reefs are found in the warm waters of the Atlantic, Pacific, and Indian oceans.
** Microbial environment encompasses the entire world ocean.
*** Ice oceans occur at both poles.

The different realms of the ocean[72]

Next to space, time is an important dimension in the expansion of CoML. While the project itself is designed to take 10 years, its research intends to cover the past, the present and the future. This is explicated in the three overarching research questions: what lived in the oceans, what lives in the oceans and what will live in the oceans? While the various field projects study the present of marine animal life in the oceans, CoML also has an historical and a future component: the History of Marine Populations (HMAP) and the Future of Animal Populations (FMAP). HMAP is using historical and environmental archives to study the past of ocean life and human interaction with the sea. Based on historic and current information about life in the oceans FMAP is trying to develop models that can predict the future of marine animal populations. "Our goal by 2010 is to know as much about life in the oceans as we know about life on land now" (Senior scientist of the Census Ronald O'Dor on CBC News, November 12, 2003).

While the Census started out as an American initiative, it has become an international endeavour with over eighty countries participating:

> The Americans actually soon realised that they can only make a Census of Marine Life when they work together with other countries. Therefore the idea has

---

[72] Image courtesy of Ron O'Dor of the Census of Marine Life.

grown into a quite unique project for biologists, certainly unique for marine biologists and ecologists. (interview Heip, 2006)

First of all, the Census has stimulated the cooperation between the United States and Europe. The EU and NSF have come to an understanding, which means that although scientists have to find money within their own national context, they are now allowed to perform the research together. "This kind of collaboration has great additional value as people in Europe and the US have different specializations that we can now bring together which gives us new insights" (interview Pierrot, 2007). And after covering the East and West Atlantic the Census soon spread towards other regions as well: "Many countries, including India and China, have strong research programmes in marine biodiversity, which should enhance the longer term focus on Census related issues" (interview Sinclair, 2006). Global expansion has been supported by the creation of regional and national nodes in Australia, Canada, the Caribbean, China, Europe and the Indian Ocean.

Nevertheless, growth has its limits. Next to problems with reaching the ambitious global goals within ten years, the tension between the international orientation of the scientific project and the national orientation of research funding complicates global collaboration. Finding funds for marine biology research is already extremely difficult, but when it concerns international collaboration it becomes even more complicated. National funding organisations are only willing to finance their own scientists and there is no source of international funding for biodiversity research. "It is true that if you have a connection with a different country or laboratory, it is not easy to fund it. While it is very worthy to work together, so I think it's stupid" (interview Sibuet, 2006). In addition, timeframes of funding programmes can conflict with the Census timeframe. While the European Union does provide money for European collaborations, they do not directly invest in CoML as its ten-year timeframe does not fit with the shorter timeframes of the Framework Programmes. As a result, national borders and funding timeframes inhibit international collaboration.

Moreover, the Census' system of financing research with national funds makes worldwide expansion impossible today, as quite some countries that border the oceans do not have funds for ocean research, notably countries in Africa and South America. Although CoML is trying to mobilise research capacity in this regions by organising meetings and sharing experiences, this has not always led to concrete research yet: "For these countries it is not a priority. And it is quite reasonable that countries in which people are hungry or have AIDS are not going to look into the deep-sea. I completely understand this of course" (interview Heip, 2006). However, this leaves CoML with a knowledge gap as this problem cannot be solved by simply letting western countries perform the research: "For us that would be very expensive and most importantly these people won't accept it. It is legally and morally almost impossible. So these coastal areas with high biodiversity are for a large part out of the Census reach" (idem). Problems of expansion of ocean research – concentrating on geographical and intellectual property – are currently on the political agenda, although they are difficult to solve as there is no international governance structure that can effectively solve these issues. As a result, the Census does not manage to become truly global and remains a bit fragmented underneath its global surface. Or, in other words: the map of life in the oceans continues to show blank spots.

Technology development

As technologies are crucial for explorations of the sea and the project's initiators are technologically orientated, technology plays an important role within the Census of Marine Life and transforms research practices and visions of life. Building on the history of ocean research, the Census makes use of the most advanced technologies and also develops them within the special technology working group. Technologies can be related to various research practices and stages. For transportation to the research area the research vessel is the most important technology. This makes marine science an expensive business as it does not only cost about 60 million dollar nowadays to acquire a vessel but also the operation costs are substantial, estimated at about 1 euro a second to have one in operation, including the expenses of personnel.[73] In addition, helicopters and planes are used, for instance to access remote areas or to study whales. For underwater exploration several vehicles are used: manned submersibles, remotely operated vehicles (ROV's), autonomous underwater vehicles (AUV's) and Deep-Towed Vehicles (DTV's) which are towed behind the research vessels.

Next to technologies for transport, the Census employs technologies for observing, counting, collecting and studying movement. For the observation and counting of life in the oceans acoustic and optical technologies are used. Acoustic technology entails sonar and echo that can also be used to study the structure of the ocean floor. Optical technologies include cameras, videos,

---

[73] According to the project officer from the European part of the Census, Dr Bhavani Narayanaswamy (personal communication, August 2006).

lasers and satellites, next to the newest versions of the good old microscope. The collection of samples takes place with the help of (traditional) fishnets, trawlers, sledges, bottles, traps and by hand. In addition, submersibles are able to bring material to the ocean surface by using their electronic arms. Finally, the movements of fish are studied with the help of fishnets, satellites, sonar, echo and the tagging of fish. For example, on the website of the TOPP project (Tagging of Pacific Predators) the movements of tagged predators such as sharks, turtles and elephant seals can be followed.

Within the research projects scientists are experimenting with the use of technologies: "It is really good that attention is given to technology. On the one hand attention is given to technologies and expertise that is already available within the project, and on the other hand new opportunities are explored" (interview Pierrot, 2006). To illustrate, the Mar-Eco expedition has used a whole range of technologies to explore their possibilities, from videosystems, ROV's to a range of different nets with various mesh sizes. The use of these new technologies makes new observations possible. For instance, with ROV's scientists can see animals in their habitat, which informs about the behaviour of species:

> Some animals are very small, but when you observe them underwater it turns out that they are able to produce a large net with which they can catch their food. However, you will never be able to see this when you catch them, as you then only see the basic body. So we only recently discovered this phenomenon with the new technologies. (idem)

As these animals eat much more with their net then expected on the basis of their body, they play a more important role in the ecosystem than thought previously. As a result, the use of new technologies make new processes visible for scientists: new information about species and their behaviour as well as new insights about food webs and the environment of species.

Next to new visions of life, the use of new technologies transform research configurations, as the spatiality of the research situation, the place of action and the area of attention changes (Lynch, 1991). Overall, current technology development in marine biology is causing a shift towards the detachment of the research subject (Ballard & Hively, 2000). "I think manned submersibles have had their time (...) We do more with ROV's and AUV's now. They can be operated from a distance, are safer and more efficient, as it takes three to five hours to descend with Alvin" (interview Heip, 2006). Instead of scientists going to their research subject technologies like unmanned submersibles and satellites and enable the performing of research from a distance. Satellite remote sensing plays an important role in the measuring of physical properties of research sites, like temperature, depth, current speed and salinity. "Satellite technologies have been a big breakthrough. Instead of point measurements on different spots you are now able to get a synoptic image in real-time, a large overview" (interview Pierrot, 2007). So instead of measuring for instance temperature at the specific research site, scientists are now able to almost immediately see the temperature of the larger waters that are surrounding them via satellite technologies.

Reinventing taxonomy

The transformation of research practices in interaction with developments in technology can also be seen in the case of taxonomy: the identification of species that is fundamental to the natural history way of knowing. Although it is a crucial practice within the Census and biology at large, taxonomy is experiencing hard times, as it is extremely difficult to find funding for taxonomic research nowadays (Bowker, 2006; Kwa, 2005). This is due to the non-charismatic character of taxonomy and the lack of fashionable technologies in identification practices. In addition, attention in biology has shifted from organisms to processes. Next to the preservation of species collections, especially the funding of scientists constitutes a problem: "The major bottleneck is money to hire people. You can get money for an expedition to collect material, but then you don't have people to do the identification" (interview Pierrot, 2007).[74] Because of this lack of funding for taxonomy, young people have to leave the field or decide to specialise in another branch of biology. So taxonomists have become an endangered species. As a result the Census focuses on the development of technologies to determine species: "They explore if there are other possibilities than the traditional labour intensive determination using a microscope" (interview Pierrot, 2007).

As genetic knowledge has the potential to fundamentally change the identification procedure, the integration of genetic technologies within taxonomic practices is now an important issue within the Census and taxonomy at large. The Census has set up a DNA working group and gave birth to the barcoding of life initiative.[75] In analogy with using barcodes to identify manufactured goods, the DNA barcode initiative wants to enable the identification

---

[74] Annelies Pierrot-Bults is a taxonomist who specialised in plankton, or, more specifically, 'Chaetognatha', which are known as 'Arrow Worms'.

[75] For more information see the website of the Barcode Marine Life Initiative and the broader Barcode of Life Data Systems (BOLD) website. Retrieved August 23, 2007 from http://www.barcodingmarinelife.org and http://www.barcodinglife.com.

of species by reading a short stretch of its genetic code: "A remarkably short DNA sequence should contain more than enough information to distinguish 10 or even 100 million species" (Stoeckle et al., 2003). Instead of looking at the whole organism, (part of) an organism is sent to the laboratory where they sequence a uniform target gene. The Census also developed a kit that researchers can use in the field to help them with the identification. This new technique is envisioned to have broad scientific applications, for example in biodiversity surveys, in identification of eggs, immature forms, stomach contents and excreta, while it may also contribute to more understanding of the evolutionary history of life on earth (Holmes, 2004).

On the one hand the use of DNA to identify species indeed enhances taxonomic practice. Molecular technologies enable the identification of species that could not be identified by traditional taxonomic methods. This especially concerns micro-organisms and living creatures from the deep sea. Genomics renders visible the enormous diversity of ocean micro-organisms that account for more than 98 percent of the oceanic biomass. Without the sequencing techniques it was impossible to see how many different kinds of bacteria a sample of seawater contains, because about 20,000 bacteria just cannot be distinguished from each other with a microscope and they cannot be grown: "So they nowadays use a DNA probe to look at the biodiversity" (interview Pierrot, 2007).[76] In addition, the development of genetic technologies helps to identify deep-sea creatures, which often cannot be brought to the surface without far-reaching damage as a result of changes in pressure.[77] This used to be a huge problem when trying to identify and classify deep-sea organisms, but now the genetic material gives information to identify the organism. Moreover, the information derived with genomic technologies also plays an important role in determining the relation between different species:

---

[76] CoML's International Census of Marine Microbes (ICoMM) is entirely devoted to the study of micro-organisms. The programme wants "to promote an agenda and an environment that will accelerate discovery, understanding, and awareness of the global significance of marine microbes" and developed the database MICROBIS. Retrieved August 23, 2007 from http://icomm.mbl.edu Retrieved August 23, 2007, from http://icomm.mbl.edu See also the press release 'Ocean Microbe Census Discovers Diverse World of Rare Bacteria', issued on July 31, 2006. Retrieved August 23, 2007, from http://www.coml.org/medres/microbe2006/CoML_ICOMM%20-Public_Release_07-31-06.pdf

[77] According to Peter Burkill from Southampton Oceanography Centre in his lecture 'Marine Biodiversity – the challenge of opening black-boxes' during the symposium 'Biodiversity in a changing world' at the KNAW in Amsterdam, June 3, 2004.

Genetic characteristics of a group of animals are now easier to determine. Before genomics technologies emerged, you could look at relationships between proteins with technologies that are already about 30 years old, but now you can look at genes. This is possible since 10 years and it already delivers spectacular results. For example in plants, like seagrass or mushrooms, it is now possible to bring various individuals back to one individual. Despite the variety they turn out to be clones of just one plant. This can only be made visible with genetic technologies. (interview Heip, 2006)

In this way genetic technologies make it possible to identify new species and the relationship between species.

On the other hand the use of genetic technologies does not really replace old-fashioned taxonomy. Although the website of CoML states that the barcoding method is "a more accurate way to identify organisms than traditional methods as it is independent of an individual taxonomist's opinion,"[78] things are a bit more complicated in practice. First of all, making the barcode system requires taxonomic expertise. All species have to be carefully identified by taxonomists before they can be barcoded. In addition, the barcode system is still in development and does not always work: "For some fishes it works and for others it doesn't" (interview Sibuet, 2006). Scientists are still exploring the reach of the new methodology. They try to use the piece of gene called cytooxidase 1 (CO1) to identify all the different species. Although this works quite well for the higher organisms, it does not work for lower organisms like plankton:

We are still exploring if it will work out (…) But to be honest we do CO2 instead of CO1 as this works better, but we are not certain if this is the answer. It works until now for us. But we try them both, to see what works and what does not. (interview Pierrot, 2007)

Finally, performing genetic tests is quite expensive. This means that in the case of many different organisms or the counting of species, morphology remains important for identification. As a consequence, scientists can use barcoding to check results or to assist identifying difficult cases, but it does not replace taxonomists and taxonomists basically view genetic technologies as an addition to morphology.

The transformations in taxonomic research practice have led to debate between the old and the younger generation of taxonomists. Prominent critique

---

[78] From the section 'Molecular techniques' at the CoML website. Retrieved August 23, 2007 from http://www.coml.org/edu/tech/identify/m-tech.htm

on the Census comes from traditional taxonomists who criticise the project's taxonomic practices. The renowned, retired researcher Alan Longhurst states that the Census neglects the amount of work that should be involved in proper taxonomy and that the project lacks reflexivity on the concept of species:

> My criticism is very simple. The original concept was planned by people having absolutely no experience in systematics and taxonomy, nor any understanding of the nature of 'species' and who believed that all that was necessary to perform a marine census was to go out, collect and if what you found appeared to be new, then it was a 'new species' to add to the census. They did not understand that to properly classify just one genus of copepods might require a 250 page monograph (see Frost & Fleminger, 1968 for an example). Worse than that, they did not even understand that the concept of a species, the basic unit of their census, is undefinable and therefore uncountable. There will be as many species in the ocean as we find it convenient to recognise.[79]

However, according to taxonomist Pierrot this critique comes from an image of the Census that does not coincide with research practice in which taxonomists are involved and the identification of species is taken seriously (interview Pierrot, 2006).

Building a new information infrastructure

In natural history collaboration, data about species are the main result of research. The way in which data are assembled, standardised, integrated and stored is crucial, not only for the research practice, but also for the future outlook on life (Bowker, 2006; Hine, 2006). Developments in information technologies have transformed the way in which data are stored, creating new memory practices. This also becomes apparent in the Census of Marine Life that is developing its own database called OBIS, which stands for Ocean Biogeographic Information System.

> Data on marine species can be as widely scattered and hard to locate as giant squid. Simplifying the search, the Ocean Biogeographic Information System (OBIS) lets you trawl 12 marine databases for collection records. (Leslie, 2002: 1685)

More specifically, OBIS combines two types of information: information on living organisms (taxonomic databases) and geographical information (GIS),

---

[79] Personal communication with Longhurst, October 5, 2006.

displaying where species have been found. OBIS has performed an important role in the formation of the collaboration, and it collects the various research results. The socio-technical connections that make up OBIS integrate the diverse research projects and underpins the collaboration that investigates life in the oceans.[80]

The idea for OBIS developed in October 1997 during a Benthic Census Meeting, leading to the establishment of a prototype website at Rutgers University (Grassle, 1997). In a series of following annual workshops OBIS took on its current shape. "The OBIS community can think of itself as a federation, loosely tied together group that aggress standards that also maintain a high degree of autonomy, especially with regard to already existing and developing data systems."[81] In 2001, the international organisation of OBIS was established, existing of an International Committee chaired by Mark Costello from New Zealand and a secretariat at Rutgers University headed by Grassle.[82] In addition, OBIS established Regional OBIS Nodes (RON) all over the world, several scientific working groups and a Science Board for quality insurance. The project is all but finished and constantly evolving and expanding. In 2006 OBIS surpassed the number of 10 million records that are freely accessible.[83]

OBIS is the backbone of the Census. It not only gathers research results, but its development also shows how its origin is interconnected with the devel-

---

[80] This section on OBIS is based on primary and secondary literature on OBIS, including a special issue on OBIS in *Oceanography* (2000), volume 13, number 3. In addition, I have attended a session on data management in marine science during the ICES Annual Science Meeting: 'Environmental and fisheries data management, access, and integration' on September 22, 2006 in Maastricht. More information about OBIS can be found at the old and the new OBIS website. Retrieved August 23, 2007 from http://marine.rutgers.edu/OBIS and http://www.iobis.org.

[81] Workshop on the Ocean Biogeographical Information System (OBIS) in Washington DC. Report of working group 1 – system design and management of OBIS, 1999, p. 1.

[82] Report from the 'OBIS International Committee Meeting' at Rutgers University, New Brunswick, USA on August 29-30, 2001. Retrieved August 23, 2007 from http://www.iobis.org/-publications/2001Report.pdf

[83] For an impression of how OBIS works: its main page basically offers two search possibilities: "search by name" or "search by geography". When entering "goldfish" this results in two different hits: the ostychis japonicus or sea gold fish and the upeneus moluccensis or goldband goldfish. Following one of the species brings you to all kinds of different information on the particular fish, including taxonomic information, characteristics, information on its discovery and a picture. OBIS also shows a map that indicates where the fish can be found. Furthermore OBIS refers to habitat/ecosystem information websites and provides a list of resources where the information is derived from. Moreover, there is a list of additional information linking directly to the specific fish in barcode of life, GenBank, Catalogue of Life, ITIS (integrative taxonomical information system), Google images, Google scholar and uBio. (This description is based on the examination of the OBIS portal in October 2006. Retrieved from http://www.iobis.nl)

opment of the wider Census program. The current international OBIS portal gradually evolved, together with the building of the Census. The technical, scientific and management components of OBIS have been subject to defining and redefining during meetings and the building of different versions of the websites. What is striking in particular regarding the development of the database is the gradual professionalising of the initiative and the careful way in which the project has been integrated in existing standards, databases and organisations. Alongside the building of the technological infrastructure, there has been ongoing discussion on what OBIS entails and how it should be integrated in the existing data community. Discussions cover the function of the database and its relation to other databases, including the standards used and how they relate to other standards.[84] Moreover, there has been reflection on how users perceive the system. From its start, the initiative has constantly been promoted in the scientific community: "everybody knows it" (interview Heip, 2006).

Within marine biology and marine research at large, the creation of OBIS is an important step forward. The deputy director of the British Oceanographic Data Centre, Lesley Rickards, claims that standardisation and integration of data have to be pursued to make research data known and compatible.[85] Although there is already a huge proliferation of databases with ocean data that are increasingly linked, more efficient use of existing data is envisioned to save additional field research efforts. However, the more efficient use of data will require collaboration and "a culture change that has to start at home" (Rickards, 2006). Rickards argues that standardisation, quality control and the use of metadata are the most important issues in contemporary marine science that will shape the future of ocean research in an important way. Her ideal is a data ATM, in analogy with the cash withdrawal machines: one node that gives access to all the data on the ocean and OBIS is partly realising this dream already.

The creation of OBIS transforms current research practices, but it is also envisioned to shape the future of research. Its builders present their database as

---

[84] On the complex process of constructing and coordinating standards, see also Bowker & Star (1999) and Schueler et al. (2008).

[85] The history of databases in ocean research is part of the assembling of geophysical data, with a crucial role for the International Council of Scientific Unions that established the World Data Center (WDC) system in the context of the International Geophysical Year (1957-1958). Nowadays the WDC system still exists, dispersed over Europe, Russia, Japan, India, China, Australia, and the United States. One of them is especially devoted to ocean research and operated by the U.S. National Oceanographic Data Center (NODC). Retrieved August 2008 from http://www.ngdc.noaa.gov/wdc; http://www.nodc.noaa.gov/General/NODC-dataexch/NODC-wdca.html.

a new approach, called Ocean Biodiversity Informatics (OBI), using information technologies to manage marine biodiversity information: "capturing, storing, searching for, retrieving, visualising, mapping, modelling, analysing and publishing data" (Vanden Berghe et al., 2006). OBI comes with the promise of new perspectives on marine life and support for the analysis of global phenomena – including climate change and over-fishing – and changes in ecosystems. As such it would mark the beginning of "a new era in biological research and management that is revolutionising the way we approach marine biodiversity research" (idem). Ausubel hopes that OBIS and the general availability of information from the CoML will allow many more people throughout the world to 'explore' marine life and gain insights in it on their own.[86] The future of OBIS and its actual use will reveal whether these high expectations are warranted indeed.

Current changes in biodiversity science are not only changing work practices of scientists but also determine our future outlook on life. In his research on memory practices in the sciences Geoffrey Bowker (2006) explains that databases not only serve as a way to structure and store research data, but they also play an important role in the shaping of scientific disciplines. He explains how databases reflect the ways in which their creators conceive of the world they live in, as well as the ways in which they order this world. For instance, during a database's construction it is decided what is put in and what is left out and how things are combined. Thereby the builders determine the way in which the world is seen through the database and, as a consequence, they shape our outlook on the world. This dynamic is well illustrated by OBIS. With the integration of taxonomic and geographic data, its builders transform research on biodiversity as the distribution of species becomes visible. But plants, for example, are not included in the Census research, which implies that this form of ocean life remains hidden from view. Moreover, by shaping our view of the world databases have an important impact on actions, for instance through policy development based on information in databases. In the case of OBIS, this is visible in models of the development of ocean life that are used for policy purposes.

Modelling the future

While natural history research has always served as a basis for learning and theorising about the development of life, this mainly concerned the history of

---

[86] Personal communication with Jesse Ausubel, January 25, 2009.

life. In contrast, the Census of Marine Life explicitly wants to learn about the future of ocean life. While it is already difficult to get a complete overview of the current state of life, predicting the future is even more complicated. The Future of Marine Animals Project (FMAP) uses models for composing pictures of the future of marine life:

> The Future of Marine Animal Populations is a network of scientists that are using statistical models to make predictions about animal life in the oceans of the future. We focus on changes driven by the fishing industry and climate change – both of these are altering the nature of marine ecosystems.[87]

FMAP develops and uses models to interpret historical data, design field studies, synthesize data and make predictions about the oceans of the future. The project grew out of a workshop held in Canada in 2002 with representatives of the wider Census, after which in 2003 FMAP was launched.

FMAP has already produced some interesting results on the development of life in the ocean. Most notably the project resulted in a prominent publication in *Science* (Worm et al., 2005) on the downward trend in the diversity of fish in the open ocean due to fishery activities. By comparing information on the number of tuna and billfish catched on a standard longline with 1000 hooks from 1952 to 1999, the authors put together an overview of the decrease of fish in the open ocean, resulting in a striking visualization of a downward trend of 50%. According to the scientists involved, this downward trend coincides with the emergence of large-scale commercial fishing and has serious consequences for marine biodiversity at large. The story was covered in newspapers worldwide, for example, in an editorial of the *Washington Post* which focuses especially on the implications of the research:

> These results are particularly disturbing because they deal with the open ocean, not with coastal waters, where depletion of fish species was already well established. A decline of species diversity could make oceanic ecosystems more vulnerable to climate change and other environmental shifts. The good news in the study is that a few hot spots remain -- though they are dramatically less vibrant than they were. One of these is off the southeastern coast of the United States. Another is south of Hawaii. These areas desperately need protection. More broadly, commercial fishing needs to be brought down to levels that will be, in the long-term, sustainable and will permit whatever recovery of species diversity is still possible, (Editorial *Washington Post*, August 13, 2005: A20)

---

[87] 'What will live in the oceans?' Retrieved August 23, 2007 from: http://www.fmap.ca

As this example underscores, the modelling efforts of Census already give rise to policy discussions.

This is not to deny that Census scientists are struggling to fulfil their promise to predict the future of ocean life because the modelling of life in the oceans proves to be a real challenge.[88] For one thing, the modelling effort within the Census still involves a relatively small effort. Although Carlo Heip, member of the Scientific Steering Committee, acknowledges the need to develop models and the impact of the first models of the Census, he has also voiced criticism of the project's modelling part:

> My problem is that FMAP is completely in the hands of one single group in Halifax and they actually are not real modellers but statisticians. So they are only looking at statistics of fisheries data and the complete biodiversity is not dealt with. (interview Heip, 2006)

According to Heip, statisticians base their effort on numbers from the past and extrapolate them to the future, while a model describes a system and tries to explain the behaviour of a system based on the processes that take place within this system. In order to provide the project's future dimension more substance an effort is made to broaden the modelling part of the Census.

Another important bottleneck that the Census scientists need to come to terms with is the difficulty of modelling the future of life in the oceans. First of all, they still do not have a proper picture of past and present ocean life, so how then design a model of its future? Moreover, they experience that models cannot handle the complexity and unpredictability of ecosystems. Complexity is a problem as models can only contain a limited number of state variables, while ecosystems contain enormous amounts of species: "When they are all put in the model as state variable, they all interact with each other and the predictive value of the models becomes zero" (interview Heip, 2006). In addition, Pierrot illustrates how ecosystems are unpredictable with a study on the effect of the rise of temperature on plankton biodiversity. This process can be studied in real life when a large area of cold water becomes isolated and flows into warmer water where it slowly warms up:

> There you can actually see what happens to species when the water becomes warmer. Although most species just cannot take it and die, you first you see

---

[88] For the difficulty of modelling fish, see Finlayson (1994) and the forthcoming PhD thesis of Diego de la Hoz del Hoyo 'Mobilizing fishy modelling across policies for sustainable fisheries', Edinburgh University.

very strange things happening. (…) Of course you can predict some things a little, but not entirely. (interview Pierrot, 2007)

If the increasing water temperature of the North Sea causes a shift in species, this does not mean that the Netherlands is getting the same ecosystem as Portugal currently has and that Dutch fishermen will start fishing for sardines: "It is not a one to one relationship. It just does not work like that" (idem). For this reason, CoML scientists need to explore ways of dealing with the modelling of ecosystems in the oceans.

Applying marine biology

While the application of research is not the primary goal of natural history research, the Census scientists experience a clear shift in research policy from fundamental towards applied research. Although the Sloan Foundation recognises the value of fundamental research into life in the oceans and supports it, other funding sources emphasise the application of research. When writing proposals to apply for funding to perform the research parts of the project, scientists often have to write a section on the application of their research. What should it deliver? Scientists have to come up with serious proposals:

> The times in which you could solve this problem with rhetoric are over. They are much more stringent now. Research has to have a clear objective (…) you can always try to sell blah-blah, but when they look through it you have a serious problem. (interview Heip, 2006)

Moreover, quite some funding organisations simply do not fund the kind of fundamental research like ocean life exploration. For instance, research funded by the EC has to contribute to the goals of the European Union: to its legislation, its policy objectives or its industry, even if the establishment of the European Research Council promises to change this in favour of fundamental research. In short, the scientists of CoML need to find ways of dealing with the requirement of application in order to get funding.

First of all, natural history research has found support in an increasing emphasis on environmental problems in our contemporary society. Carlo Heip explains that when he started his career in the 1970s, the problem of pollution was prominent and marine biology concentrated on heavy metals and their influence on living organisms and ecosystems:

> I have witnessed how Carson with *Silent Spring* [1962] and the Club of Rome had a deep impact on marine science and people in general. The idea that peo-

ple are able to destroy the planet emerged in the 1960s when I was a student. So then we had an emphasis on mechanisms: why does it work that way? But you also have to procure funding, which in the 1970s could be done in the context of pollution. (interview Heip, 2006)

In the 1990s the issue of climate change emerged and biodiversity has acquired prominence after the 1992 Convention on Biological Diversity.[89] For instance, the impact of fisheries on marine ecosystems has emerged as a central issue in the last ten years: "That seems to be most significant now, that fisheries are able to influence the open ocean, while pollution is predominantly restricted to coastal areas" (interview Heip, 2006). As becomes visible within the research and modelling practices of the Census, issues such as climate change and biodiversity shape contemporary explorations of the ocean.

In addition, research within marine life has concrete (industrial) applications. Technology development is an important contribution of marine biology, for instance in the areas of information technology, the tracking of organisms, satellite connections, online observatories and genomics. In analogy with space research, marine science helps to develop new materials, such as isolation material. However, underwater circumstances also provide knowledge about what happens with life at low levels of oxygen: oxygen hypoxia. "Although this seems not useful at first sight, you can actually study what happens if people drown or suffer from carbon mono-oxide poisoning" (interview Pierrot, 2007). Another good example of application is the use of newly discovered marine microbes. The US Department of Energy, for instance, expects microbes to ultimately solve 'challenges' concerning energy production, global climate change mitigation and environmental cleanup.[90]

For the application of research, funding organisations often stimulate collaboration with industry. In the case of marine science, this may involve an array of companies and business activities, ranging from aquaculture or fisheries to instrument makers and consultancy. Likewise, the pharmaceutical industry is a good partner for collaboration, even though Heip – who established Euro-CoML – points out that the European Union gives priority to funding collaboration with small to medium-sized companies, rather than the large pharmaceutical corporations. The drawback of working with smaller companies, however, is that often the collaborative efforts do not deliver any profits. In the development of instruments, for example, the number of sales is always very small,

---

[89] Retrieved May 23, 2008 from http://www.cbd.int/
[90] "Many challenges have possible solutions in the invisible biological world of microbes" The DoE genomes to life program, Retrieved May 23 from http://doegenomestolife.org.

because the highly specialised instruments do not have a large pool of potential users. "These companies often cannot permit to send an employee to a three-day conference to talk with scientists because it hardly has any direct benefits. We experience this as a problem in collaboration" (interview Heip, 2006). In collaborations with smaller industries, then, not only the research but also the costs of the company need to be covered.

The political emphasis on application leads to quite some frustrations by Census scientists:

> To be honest, I think research should not always be useful in the first place. Of course you have to account for what you do towards society. However, use should not always be equalled with money. I find that a very narrow way of thinking (…) I always think that it is good to know how the oceans work, you want to know how the system works and why some systems are more produc-tive then other systems, and what happens if the system changes. (interview Pierrot, 2007)

To voice the importance of ocean exploration, marine scientists developed a tradition of lobbying to convince governments of the importance of their research. In addition, scientists have become creative in procuring funds. They have found ways to perform research with money that was officially not in-tended for fundamental research: "Scientists are smart of course, so lots of money is actually used for fundamental research even if it was not originally intended for it" (interview Heip, 2006). Today, the scientists of CoML increas-ingly rely on the recipe of Sloan: 'capture the imagination' of the public.

Showing the public

Although the underwater world has been a public attraction since the emer-gence of (public) aquariums, Jules Verne's *Vingt mille lieues sous les mers* (1870) and the movies of Jacques Cousteau, the Census is providing a new impetus to the public's awareness of life in the oceans. "The oceans, like the heavens, offer a preferred route to increasing public understanding of the world in which we live, and of science" (Ausubel, 1997: 1). Public communication sets the Census of Marine Life apart from other projects in marine science and thereby the project reflects the current trend towards the embedding of science in society that is a central characteristic of big biology.

On behalf of the Sloan Foundation, Jesse Ausubel has been the driving force behind this special approach from the beginning of the project. He explains the Sloan Foundation's view of science as a privilege, not a right, and that researchers in general should share what they do with the society that supports and permits them to do it and that they should listen carefully to the public. "Sloan does not think of 'public relations'. Sloan does seek to advance both the scientists' understanding of the public and the public understanding of science".[91] Ausubel – "who invests lots of time and energy and never seems to get tired to talk to everybody about the project" (interview Pierrot, 2007) – is backed by the international secretariat of the Census that developed a public relations policy. The secretariat forms the central public communication node, but in addition individual projects are required to pay attention to interaction with the public.

First if all, the Census makes sure that its results are communicated to a wider public via its press policy:

> They have an excellent press policy. They perfectly communicate the high-lights of the project. When the Census issues a press release, it appears in newspapers all over the world. (interview Heip, 2006)

A search in a selection of major international newspapers already delivers 210 articles about 'Census of Marine Life' over a period of one year.[92] Also the website is an important part of the public relation strategy of CoML. It has a main portal which gives general information on the Census and an introduction to its different components. In addition, each project has its own website, on which detailed information is given on research plans, activities and output.

On top of making public communication daily business, various special initiatives have been developed, including a book and an expedition called "Deeper than Light", which now travels the globe. In addition, a movie is made. The Galatée Films Ocean Project is headed by the renowned French film director Jacques Perrin who already made successful movies about monkeys, insects and birds. Perrin, who is now filming life in the oceans in collaboration with Census scientists, has claimed:

> *Oceans* hopes to be a global project, like Census of Marine Life. Filming will last 30 months, with three crews dispersed around the world. We want to take our time and not rush the natural order of things. We want to be at the service of nature, listen to her.[93]

---

[91] Personal communication with Jesse Ausubel, January 25, 2009.

[92] Using the international news database of LexisNexis over the period of 23 August 2006 till 22 August 2007.

[93] Perrin is cited in 'The Ocean Project Press KIT' produced by Galatée Films in 2006. Retrieved August 23, 2007 from http://www.coml.org/medres/galatee/Oceans_Project.pdf

The release is scheduled for the fall of 2009 and the movie will be staged in cinemas all over the world.

Last but not least, the Census invests time and money in education. The website has a special section which explains its research and technology to laypeople. A nice example of an educational program is the project 'cruise in your classroom' which is a website diary of cruises that are also made in the context of the Census.[94] The website shows what it means to go on a cruise and scientists tell you about their life at sea. For example, scientist Caren Braby who took part in the Easter Microplate Expedition which was part of the Census project on hydrothermal vents, tells in her logbook about her first meeting with Alvin:

> We are still motoring towards our first site at 38°S. We expect to arrive on the evening of March 20, so our first dive should be Monday, March 21. My observations upon "meeting *Alvin*" are here (hyperlink to text below). During my orientation it was so exciting just to sit in *Alvin*—even with the hatch open and even with *Alvin* locked down in the hangar. I was listening carefully to the pilot giving us our orientation but had time to daydream about being a mile and a half deep in the Pacific Ocean, looking for adventure. It's an amazing machine and I feel so lucky to have a first hand view of its capabilities.[95]

The Census projects and scientists are involved in educational activities to make children aware of the importance of our living environment and stimulate them to choose a career in science.

New natural history

The Census of Marine Life is part of a long tradition of scientific collaboration in natural history. In this chapter I examined the Census in the context of what current natural history collaboration entails. I also explored differences and similarities between old natural history expeditions and contemporary explorations of life on earth. First of all, continuity can be seen in the character of the natural history collaborations. As Pickstone (2000) has argued, the natural history way of knowing continues to exist alongside newer ways of knowing. In this respect Rip (2001) emphasised the continuity of the scientific practice of

---

[94] Retrieved August 23, 2007 from: http://www.mbari.org/expeditions/index.htm
[95] Braby on 16 March, 2006. The logbook can be found at the website of the Easter Microplate Expedition. Retrieved August 23, 2007 from: http://www.mbari.org/expeditions/Easter-Microplate/March16.htm

measuring, mapping and modelling. This continuity is also implied in the very design of the Census project. For one thing, the scientists named their project 'Census': it is about counting and mapping what populates the sea. And during one of the initial meetings of the Census, the project was presented as part of the exploration of the world: "The age of discovery is not over. Indeed, the voyages of discovery open to Charles Darwin, Captain Cook, and the explorers of Linnaeus' century are very much open to the voyagers of 2000 and beyond" (Ausubel, 1999b: 4).

Although continuities with the past can be observed, my analysis of the Census of Marine Life illustrates the reinvention of natural history collaboration as it shows particular changes that take place in this style of collaboration. To start, the scale and scope of natural history is becoming ever larger. Nowadays, the exploration of the earth takes place in the context of a globalising world that is predominantly shaped by knowledge and technology (Bijker, 2006; Castells, 1996; Held, 1999). With the participation of more then 2000 scientists from more then 80 countries, the Census research attempts to cover all the worlds' oceans, broadening the scope of research geographically. As a result, mapping has basically become a global effort.

Next to globalisation, taxonomic research, a vital part of natural history, is transforming fundamentally. Although after the 18th century natural history became subordinate to the latest developments in biology, involving analysis and experimentation within the closed world of the laboratory (Pickstone, 2000), these old and new ways of knowing are increasingly integrated again. The new styles of knowing gave rise to molecular biology and genomics, which is transforming taxonomic research. Where taxonomists traditionally use morphology to identify species, now a shift is taking place towards genetic identification of species. This broadens the biological scope of the research, including the animals of the deep-sea and the world of micro-organisms.

In addition, the integration and contextualisation of knowledge can be observed. An important criticism on the Census is its focus on identification and cataloguing without explanation: "For example in the US they say 'oh the project that wants to count all the animals in the sea'. You know, that kind of degrading remarks. They don't consider it science but just exploration" (interview Heip, 2006). However, the Census actually shows how present-day natural history is not only about counting life. Although identification and cataloguing of species is still central, this is increasingly presented as a starting point for the creation of new knowledge through the integration of data. The inventory of ocean life is a tool that can be used in further research on the interaction between species and their environment: "We have to start with an inventory of good quality. In addition, you may then really focus on questions to explain relationships within biology" (interview Sibuet, 2006). This increasing focus on ecosystems means the integration of information about life and geography, which becomes visible in OBIS and modelling initiatives that contextualise knowledge about life.

Finally, technological development and new relationships between science and society transform research practices. My analysis of the Census has shown how the development of new technologies is part of changing research configurations that bring new visions of life. This cannot only be seen in the transformation of taxonomic practices through genetic technologies, but also in the widening of observation through satellite technology and the building of the new information infrastructure OBIS, creating a new outlook on life in the oceans. Current developments in the relationship between science and society are reflected in increasing attention to public communication and the application of marine research.

In sum, the analysis of the Census of Marine Life has shown how recent transformations in biology and the place of science in society are not only shaping new collaborations in biology, such as the Human Genome Project, but also transform the more traditional style of collaboration known as natural history in particular ways. While Rip talks about the 'sophistication' of natural science practices, as being mainly due to developments in information and communication technologies, my analysis of the Census established that natural history has transformed more fundamentally and that transformations go beyond technology development and also incorporate social developments. Contemporary developments in science and its organisation become integrated in the traditional natural history style of research. As a result, *new* natural history collaboration integrates traditional practices with recent developments in biology and society, transforming the way in which life is measured, mapped and modelled. [96]

Moreover, my analysis of the Census shows how the new ways of doing natural history come with their own particular problems. While the process that Rip (2001) calls 'sophistication' implies that measuring, mapping and modelling practices are now more advanced and maybe even more effective, my research shows some major problems in today's natural history. For instance, the use of genomics technologies for identifying species does not seem to solve the shortage of taxonomists, but gives rise to discussions about proper taxonomy instead. In addition, tensions between international research and national funding structures are an important bottleneck for research, as is true of the lack of

---

[96] This is in line with a recent article of Pickstone in which he explicitly stresses the combination and interaction between different ways of knowing (Picktone, 2007: 491).

international governance structures geared to stimulating and regulating international ocean research. This has caused the limits of growth in natural history collaborations to become apparent: not all countries participate and not all species are catalogued. And finally, the Census of Marine Life struggles with integrating the research materials and building global models. However, the project has also underscored the remarkable resilience of big natural history, as the project seeks to extend itself into the future eventually to accomplish its goals.

CHAPTER 4

Growing a cell *in silico*
Constructing collaboration

Visions of the cell appeared in the 17th century when Robert Hooke observed the basic structure of cork through a microscope and noticed spaces that were similar to the small rooms monks used to live in, so-called 'cella' (Sloterdijk, 2005; Hooke, 1665/1987). Since then, the cell has become an important unit of study leading to the establishment of cytology – currently known as cell biology – and an increasing understanding of the cell (Maienschein, 1991).

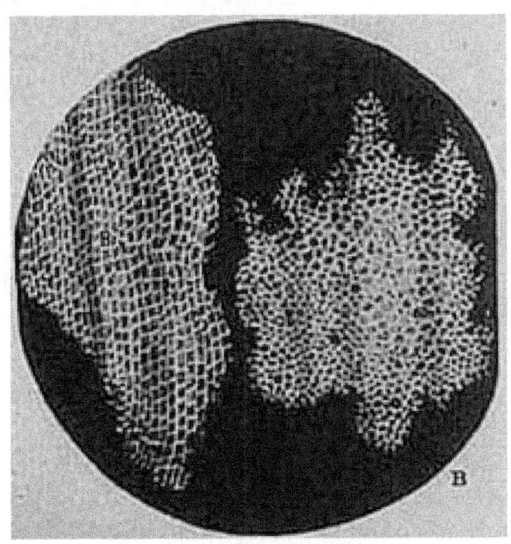

Cells in Micrographia[97]

---

[97] Derived from the book written by Hooke, 1665: Schem:XI, Fig:1

More than 450 years after its discovery, scientists are now rebuilding a cell *in silico*: a replica of a living cell in a computer.

> A Silicon Cell is a precise replica of (part of) a living cell. It is based on ex-
> perimentally determined rate laws and parameter values, *i.e.* only on data, not
> on fitted values or assumptions. It merely calculates the system biology impli-
> cations of the molecular properties that are already known.[98]

The Silicon Cell does not resemble Hooke's monastery room at all, nor does this cell look like common textbook images of cells. It consists of an enormous amount of letters, numbers and moving lineages forming graphs on a screen representing specific cellular processes.

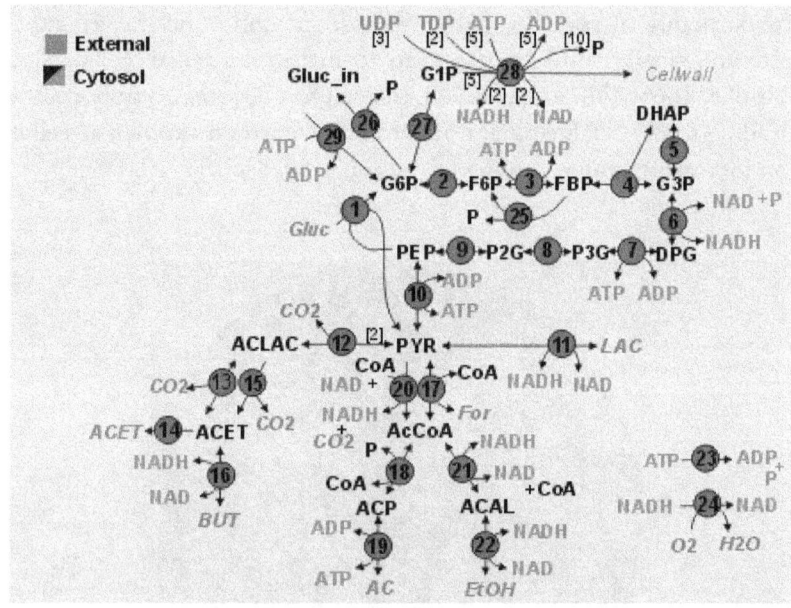

Part of a silicon cell[99]

The idea of a Silicon Cell involves the convergence of two recent developments in biology. First, investigations into the molecular basis of life have resulted in a wealth of information on the cell, its constituents and processes. Secondly, the

---

[98] Retrieved April 30, 2007 from http://www.siliconcell.net.

[99] Image from the website JWS Online: Online Cellular Systems Modelling. Retrieved June 21, 2007 from http://www.jjj.bio.vu.nl

informational turn in biology that came together with the increasing use of information technologies, now make the integration of all this information on the cell in a computer model feasible. The virtual model of a living cell is envisioned to mimic all the processes of an *in vivo* cell and to replace living cells in scientific experiments, making research faster, less complicated and more reliable. The Silicon Cell relates to initiatives to build even more complex entities of life *in silico*, like a virtual plant, a virtual heart, a complete virtual human and even the replication of whole ecosystems. The development of these computer models cause a shift from 'wet research' in the laboratory to 'dry research' using information technologies (Penders et al., 2008). Moreover, the modelling of organisms is part of the movement towards 'systems biology'. While genomics and other -omics research primarily focuses on the analysis of different parts of the cell, systems biology aims to make sense out of the huge amount of data through integration and contextualisation in models.

While the former chapter described how collaboration in the natural history tradition has changed nowadays, this chapter is about constructing collaboration in modern laboratory biology. About the creation of a collaboration that wants to integrate new molecular knowledge about cells, in order to create a model of the living cell in a computer. A small group of scientists working on the Silicon Cell is trying to orchestrate the European research community towards the development of a replica cell. They present the construction of a Silicon Cell as big science, as the analysis and integration of all the different parts of a complex cell requires a huge effort that will be even bigger than the Human Genome Project. However, despite the powerful image of the cell and its application in (industrial) research, the scientists behind the Silicon Cell initiative did not manage to put the actual collaboration together at the end of my investigations in 2007 and it is still not there when finishing this book at the end of 2008. The silicon cell is far from a virtual reality. What is going on? Why does it turn out to be so difficult to make biology big?

With the Silicon Cell Initiative I studied a collaboration that did not succeed. While studying failure is known to lay bare the mechanisms of a certain process and in this case shows how the making of connections can be hard, this type of research has not been done within the study of scientific collaboration yet.[100] By investigating the attempts to construct the Silicon Cell project, the difficulties of building big science are made more apparent precisely because of the explicit efforts being made to achieve it. This gives me the opportunity to explore the dynamics of the building of collaboration in the life sciences. Where

---

[100] For the importance of studying not only successful cases see Bijker & Pinch, 1987; Barnes & Bloor, 1982; Latour, 1987

the conceptual part of this book gave a macro perspective on transformations in the life sciences, and the former empirical chapter put collaboration in biology in a historical perspective by looking at transformations in natural history type of collaborations, this chapter will describe in detail the formation of collaboration. How is collaboration built, and what is the role of scientists involved? The analysis of this project is the second step in the exploration of central issues in scientific collaboration in this empirical part of the book, which will subsequently form the basis of theoretical reflection in this study's last part.

To show what it takes to build collaboration, I will use a dramaturgical perspective to sketch a rich picture of the development of the Silicon Cell Initiative (Goffman, 1973; Hilgartner, 2000; Bal et al., 2002). By analysing scientific collaboration as a performance, collaboration is seen as a complex task analogous to staging a play. A collaboration consists of actors that come together to cooperate towards a collective result. They have to build a team, manage information, construct a project story, present their project to different audiences (e.g. other scientists, government, business and public), compete for money, perform research and receive critique. Analysing collaboration as performance also makes a distinction between backstage and front stage; backstage the actual research practices takes place while front stage the scientific work is presented to the audiences. For this reason, I will show how scientific collaboration does not only mean performing research together, but also involves the presentation of collaboration. Moreover, the dramaturgical perspective reveals transforming roles of scientists.

In my analyses I distinguish three different stages – or acts – in the process of developing the Silicon Cell Initiative. The first act 'Visions of a cell in a computer' will look into the origins of the Silicon Cell Initiative. Subsequently, 'Staging the Silicon Cell' examines how the project is put forward as big science. However, as this did not result in actual funding for the initiative, a change of strategy followed consisting of a turn towards systems biology. This forms the third act. Finally, I will review the construction of large-scale collaboration. [101]

---

[101] My analysis of the Silicon Cell Initiative and the systems biology efforts discussed later in this chapter is based on literature about SiC and systems biology, consisting of websites, policy documents, meeting minutes, scientific articles and books. In addition, I have performed in depth interviews with scientists and policymakers (for an overview of interviews, see Appendix A). Moreover, I have observed meetings dedicated to systems biology (for an overview of meetings, see Appendix B).

Visions of a cell in a computer

In a letter to Robert Hooke, Newton famously used the old metaphor of 'dwarves standing on the shoulders of giants', pointing out that his ideas strongly build on the work of Descartes (Merton, 1965/1993). The shoulders became an important notion in the understanding of the development of science. Scientific ideas always build on previous investigations and it is difficult to pinpoint the emergence of a specific idea. Moreover, science in the making is quite a complex and messy process (Latour, 1987). Often only with hindsight a storyline with causal relations can be constructed, giving a linear and neat view on the complex development of science. The same complexity appears when analysing scientific collaboration, adding an organisational twist. This section discusses the development of scientific collaboration, investigating the development of the Silicon Cell Initiative. How does a scientific collaboration originate? What can be seen as the beginning of what is later defined as a research project? Do ideas form the beginning of a scientific collaboration or is it an organisational matter from the start? Sometimes, collaborators present a clear story of origin – what Knorr-Cetina (1999) calls a birth drama – but in other cases it is difficult to trace back the roots of a scientific collaboration, for the analyst as well as for the scientists involved.

When asked about the origin of the Silicon Cell Initiative, Professor Hans Westerhoff, head of the department of molecular cell physiology at the Bio-Center Amsterdam and the key person behind the initiative, points directly at PhD research of two of his former students. But then he hesitates and starts reflecting by asking: "How does something like this start?" (interview Westerhoff, 2005).[102] It turns out to be quite difficult to trace back the emergence of the initiative. However, gradually the history of the Silicon Cell is reconstructed from bits and pieces of Westerhoff's memory, complemented by the view of other participants and documentation about the initiative. It turns out that not only PhD students, but also yeast and sleeping sickness play an important role in the origin of the Silicon Cell, as well as the overall goal of finding a cure for cancer. Moreover, I will show how a scientific dispute, a crucial experiment and going public are key elements in the development of the Silicon Cell Initiative.

---

[102] Prof dr Hans V. Westerhoff is leader of the Department of Molecular Cell Physiology at the Faculty of Earth and Life Sciences of the Free University in Amsterdam. In addition, he is a group leader for systems biology at the Manchester Centre for Integrative Systems Biology (MCISB) in the Manchester Interdisciplinary BioCentre (MIB), which is part of Manchester University.

Doctoral research on three different organisms – yeast, *T.brucei* and *E.coli* – constituted the beginning of the Silicon Cell. First of all, Bas Teusink investigated sugar decomposition in a yeast cell and found oscillation as mechanism behind this process: glycolytic oscillations.[103] However, another scientist from another group came with a different mechanism which could also explain the phenomenon. This resulted in a heated academic debate, which made Westerhoff think of ways to solve the dispute: "This moment stirred me up as I thought: this is not possible. One person fully defends one mechanism by shouting and railing against other people who have a different model and actually both parties do not have a leg to stand on" (interview Westerhoff, 2005). Westerhoff figured that the only way to solve the dispute was to make a precise model of the process in the cell. The actual development of a computer model showed that a combination of the two mechanisms was in place. So, it was an attempt to solve a scientific dispute in the context of PhD research that led to the building of the first Silicon Cell.

Subsequently another doctoral student and another organism entered the stage, presenting another route towards the Silicon Cell. Barbara Bakker investigated *Trypanosoma brucei* in order to find an effective way to tackle the parasite that causes sleeping sickness.[104] Bakker's research tried to find a way to kill the parasite by inhibiting a step in its metabolism: "In other words, you have to find a place on the metabolic highway where you can put a stone so the cars cannot drive anymore" (interview Westerhoff, 2005). The challenge is to find a crucial step or enzym that can be inhibited but does not appear in the metabolism of a human cell. In this way, the drug will only kill the parasite and will leave the human cells intact.[105] The building of a model made it possible to calculate which step this needed to be: "A fantastic result" (idem). Finally, the third project of Johann Rohwer, on sugar decomposition in the bacterium *E-coli*, also successfully employed the modelling technique.[106] If these doctoral

---

[103] Bas Teusink is currently working in the bacterial genomics group at the Centre for Molecular Life Sciences at the Medical Centre of Radboud University Nijmegen. Retrieved June 5, 2008 from http://www2.cmbi.ru.nl/who-and-where/staff/35.

[104] Barbra Bakker is again part of the group of Westerhoff at the Free University in Amsterdam and is project leader of the IOP Vertical Genomics project, a collaboration between academia and industry studying cell processes in yeast (IOP Bulletin 3, 2007: 3).

[105] This process is known as differential networked-based drug design.

[106] Johann Rohwer returned to his home country South-Africa where he is part of the Triple-J Group for Molecular Cell Physiology at Stellenbosch University. Retrieved June 5, 2008 from http://www.jjj.sun.ac.za.

research projects provided the foundation for the Silicon Cell Initiative, it was a crucial experiment that solidified the potential of the Silicon Cell as the next section will explain.

Going public after a crucial experiment

While the idea of the Silicon Cell implicitly took shape within the context of the different lines of PhD research at the beginning of the 1990s, it became explicit in 1997. At that time the model definitely proved to work: "At that moment we were able to use the calculations to find a drug target for *T.brucei* and we also accidentally discovered a biological phenomenon, the functionality of something" (interview Westerhoff, 2005). More precisely, the Silicon Cell played a crucial role in the discovery of the function of the organel in *T.brucei* and yeast cells. These cells are unique because they have an organel around the enzymes that decompose sugar. By using the computer model, the scientists were able to see what happens when removing this organel: the cells explode. According to Westerhoff this experiment could not have been done in a real cell, as the removing of one part would have deranged the complete cell. Therefore the experiment proved the use of the computer model, which was a crucial step in the development of the Silicon Cell Initiative: "At that moment you have a model that works, you lay your hands on something with which you can go public" (idem).

Going public meant communication of the scientific observations and results to fellow scientists, which led to the articulation of the idea of the Silicon Cell. "You are enthusiastic about the results and you start telling stories about it" (interview Westerhoff, 2005). The stories were followed by lectures, invitations to conferences and writing results down: "And this is where you start thinking about the wider significance because that is important when you tell your story: why is it interesting for other people?" (idem). The process of presenting the research results made Westerhoff think about the Silicon Cell as a principle or practice that should be applied more widely. As a result he started to mention the Silicon Cell in articles and he constructed the Silicon Cell website together with Jacky Snoep, who had worked in Amsterdam before becoming a professor at Stellenbosch University in South-Africa.

Contextualising the research

In the presentation of research to others, the context and use of research becomes important. Work on the Silicon Cell and the interest in the functioning of cells stems from an interest in improving human health. Westerhoff spend an important part of his career working at the Dutch Cancer Research Institute[107] and kept his interest in cancer research as head of the research group at the Free University in Amsterdam: "We were actually interested in tumour cell biology and the development of drugs, but that turned out to be too difficult because we did not know the system well enough" (interview Westerhoff, 2005). 'The system' refers to the way in which the elements of a tumour cell interact. Every cell consists of various elements that independently of each other do nothing. Only in interaction these different parts make up the processes in a living cell, like the cyclical process of cell division. So when wanting to control processes in a tumour cell in order to fight the cancer, you have to figure out which enzymes can influence the processes in the cell most effectively. This can be done by modelling: the Silicon Cell.

As the modelling of cancer cells is very complex, another less complicated cell had to be chosen to begin with: "So then we started thinking about which system we could use to start constructing these precise models and that turned out to be yeast" (interview Westerhoff, 2005). The choice for baker's yeast, or *Saccharomyces cerevisiae*, was both scientifically and economically interesting. Yeast is relatively simple and well-known as it functions as a model organism in biology and it is fundamental to various important industrial processes like baking and brewing. As a result, a small group of scientists around Westerhoff started to investigate and model the mechanisms in the yeast cell – which included Teusink's research on glycolytic oscillations – as one small step towards the modelling of cancer cells.

Also Bakker's research is presented as part of the wider goal to model cancer. The parasite *T.Brucei* that figured in Bakker's research is a relatively simple organism too. Moreover, this organism was chosen because two colleagues of Westerhoff in Brussels already did extensive research into this organism.[108] Bakker's accomplishment to find a drug target for the sleeping illness by identifying the sugarintake as the step that should be inhibited, is one step in the right direction. "However, the final goal is still the tumour cell and there the problem is of course that this cell is almost similar to the cells of the host" (interview Westerhoff, 2005). This means that it is very likely that when you inhibit a step in the process of the tumorcell, you also inhibit the healthy cells.

---

[107] The Dutch Cancer Research Institute (NKI). Retrieved April 25, 2007, from http://www.nki.nl.

[108] These colleagues are Fred Opperdoes and Paul Michels who are based at the Research Unit for Tropical Diseases which is part of the Christian de Duve Institute of Cellular Pathology and the laboratory of Biochemistry, Catholic University of Louvain, located in Brussels. Retrieved April 24, 2007 from http://www.icp.be/trop.

This makes the case of cancer much more complex. In sum, the Silicon Cell is inspired by the dream to cure cancer and an ultimate aim is the building of precise models of a cancer cells to find ways to effectively inhibit vital processes in these cells.

Back to the beginning of collaboration

The early development of the Silicon Cell initiative illustrates that tracing the beginning of a scientific collaboration brings similar difficulties to detecting the origin of an idea, as both the idea and the collaboration build on a tradition of ideas and existing social ties. The idea of the Silicon Cell emerged in a tradition of cell biology, combined with recent developments that make the modelling of processes in the cell possible. Moreover, the Silicon Cell is part of an effort to improve human health, through exploring ways to cure cancer. However, when investigating the origin of collaboration, the organisational context of the research topic becomes prominent as well.

The idea of the Silicon Cell emerged within a specific organisational setting that formed the basis for later collaboration. The Silicon Cell got shape within the traditional academic research group of Westerhoff, which consisted of about eight people at that time. However, the research on yeast was embedded in the international community that studies this model organism, which is also the case for the study of the other organism *E. Coli*.[109] Moreover, the *E.Coli* project was the result of a connection between Westerhoff's group in Amsterdam and the molecular cell physiology group in Stellenbosch, South-Africa. Originally from South-Africa, Johan Rohwer and his project build upon this connection that was established by sabbaticals from Westerhoff and Professor Jannie Hofmeyer from Stellenbosch and the professorship of Jakie Snoep. Similarly, the research on *T.brucei* was done in collaboration with colleagues in Brussels. In short, the research projects that provided the basis for the idea of the Silicon Cell were embedded in different organisational networks.

In addition, the case of the Silicon Cell shows how interaction with other scientists that are not involved in the research process, takes a central place in the development of the idea of the Silicon Cell. First, the presentation of research results led to a scientific dispute that was essential for the development of the idea of a cell replica. Later, the articulation of the idea of the Silicon Cell

---

[109] For the E. coli consortium, see the 'International E.coli Alliance' and the 'E-Coli hub'. Retrieved June 21, 2007 from http://www.uni-giessen.de/~gx1052/IECA/ieca.html and http://www.ecolicommunity.org. A yeast consortium can be found at '*saccharomyces* Genome Database. Retrieved June 21, 2007 from www.yeastgenome.org.

took place when communicating research results to a broader scientific public. Within this interaction the wider relevance of the principle became clear, as well as its potential for further exploration and application. In other words, the presentation of research to scientists outside of the research group helped to shape the Silicon Cell Initiative.

In conclusion, science and its organisation cannot be separated as the development of an idea always takes place within an organisational context. When analysing the origin of a scientific collaboration, the interaction between both elements becomes relevant. The idea has to be traced back together with its organisational embedding. In other words, when looking into the origin of collaboration not only the giants, but also their homes and their connections to others have to be investigated. Although the idea of the Silicon Cell did emerge in the interaction between scientists, those relations did not have the form of a structural collaboration around the Silicon Cell yet. Only after its articulation the idea of the Silicon Cell became a starting point for further organisational activity. However, as I will explain in the next section, the process of setting up official collaboration strongly builds upon the relations that already existed in this first act.

Staging the Silicon Cell as big science

In the second phase of the Silicon Cell Initiative, the cell becomes the central theme of several organisational activities aiming to build a large-scale research project. This section describes how the initiative evolves towards the aim of building a formal collaboration. With the website as a starting point, I will discuss the Amsterdam Silicon Cell consortium and proposals for research collaborations on a Dutch and European level in which the Silicon Cell Initiative grows and is put forward as big science. The analyses of a project in development illustrates that the elaboration of ideas and organisational matters continues to be interwoven as they were in the first stage. However, I will also show that a division emerges between ideas and plans on the one hand and the realisation of those plans on the other hand. An international Silicon Cell programme that is even bigger than the Human Genome Project is still a vision.

Building a website

Together with Snoep from the group at Stellenbosch University, Westerhoff constructed a website entitled 'SiC!: The Silicon Cells' which can be found at the special domain siliconcell.net (See screen 1). The website forms the virtual heart of the initiative and exists of general information about the Silicon Cell, a separate website with the actual models and an overview of organisational activities.

SiC!: The Silicon Cells

A silicon cell is a precise replica of (part of) a living cell.   It is based on experimentally determined rate laws and parameter values, *i.e.* only on data, not on fitted values or assumptions. It merely calculates the system biology implications of the molecular properties that are already known. Silicon cell is not a package of software for simulations. The international silicon cell program thereby differs (i) from the *Virtual Cell* (www.nrcam.uchc.edu), which offer web-based complete modeling environment for cell biology that can be used to calculate what happens in cells (and is more powerful for spatial aspects) and has some actual models for the purpose of demonstrating the procedure), (ii) and from the *E-Cell* which also offers such an environment but only for down-loading to one's own computer and E-cell (http://www.e-cell.org/), (iii) from Symphony of Genomatica (http://www.genomatica.com/index.shtml , which is not freeware but also has organism and pathway specific information) in that it calculates kinetics, rather than analyzing which pathways are possible or actually used.

At present silicon cells exist for glycolysis in yeast, trypanosomes, *E. coli*, erythrocytes, EGF induced signal transduction, for histone-gene expression in early development. Most of these can be found on the ready-to-use website (also ideal for teaching purposes) pioneered by Jacky Snoep.

☐   Silicon Cell ready to use : the website with silicon cells that can be run over the web. Please contact us when interested in joining this international consortium.

☐   Extended examples and instructions for using the Silicon Cells (i.e. the website; contains questions with answers; may be used for teaching biochemistry and integrative bioinformatics courses)

☐   Amsterdam Silicon cell papers

☐   The Amsterdam Silicon Cells programme

    YSiC: the Yeast Silicon Cell, a planned European Consortium

To The   *BioSim*   page
NETWORK OF EXCELLENCE

Towards the System Biology site

Screen 1[110]

First of all, the main page presents the Silicon Cell to a wider public by giving a precise definition of the Silicon Cell and positioning it against other cell model-ling initiatives. Making a model of a cell is the central theme of different pro-jects scattered around the world: like the e-cell from the Institute for Advanced Biosciences at Keio University in Japan[111] and the Virtual Cell project in the United States.[112] In contrast to these projects, the Silicon Cell is not a package of software for simulations but a replica:

---

[110] Retrieved April 30, 2007 from http://www.siliconcell.net

[111] Retrieved May 26, 2008 from http://www.e-cell.org/ecell

[112] The Virtual cell project is funded by the NIH's National Center for Research Resources and based at the National Resource for Cell Analysis and Modeling at the University of Connecticut Health Center. Retrieved May 26, 2008 from http://www.nrcam.uchc.edu

A simulation can be a replica but in practice it often is not, as people just take a model and try to make it simulate the real situation by adjusting some parameters. That is absolutely forbidden in the case of our replica: we are not doing that. We are only allowed to change a parameter if we measure that in the experimental situation because we want to explain the behaviour of the system from the molecular behaviour. (interview Westerhoff, 2005)

So, while a simulation has to mimic the processes in a cell, a replica imitates them exactly.

Nevertheless, the use of the word replica is contested, even within the Silicon Cell Initiative, as the Silicon Cell is only a replica in the computational world and does not have the actual mass and chemistry of a real cell. Westerhoff acknowledges this argument but does not consider it relevant:

This is interesting, but not important at every moment. I just think it is a shame if people do not want to call it a replica, because if everybody just calls it a replica everybody uses the same word and although we can say that it is maybe not the best word, it is a word that makes clear that something special is going on. (interview Westerhoff, 2005)

In sum, Westerhoff employs the word replica to demarcate the Silicon Cell from other initiatives that make computer models of cells.

In addition to demarcating the Silicon Cell from similar initiatives, the website present results of research by linking to a separate website: the 'Silicon Cell ready to use: the website with Silicon Cells that can be run over the web'. When accessing this website you enter a somewhat different world which is constructed by scientists from Stellenbosch University.[113] The website has two objectives: putting the available models on the web to make them accessible for others and tempting scientists from all over the world to add new models to expand the Silicon Cell Initiative. The website contains a model database including Silicon Cells: from glycolysis in yeast, *T. brucei* and *E. coli* to red blood cells and the process of photosynthesis. As a result, the database encloses the result of research on processes in various cells and is therefore the materialisation of the Silicon Cell initiative.

Finally, the Silicon Cell website outlines the plans for the building of a Silicon Cell initiative. The Silicon Cell wants to be an open initiative "Everybody can do a small piece of the puzzle, which will together make the bigger picture. As we can never do it alone and I think nobody can" (interview

---

[113] Jakie Snoep is responsible for the content – as Mathematica Developer and database curator – and Oliver Brett is responsible for the website and systems administration.

Westerhoff, 2005). Therefore the website also explicitly wants to introduce scientists to the Silicon Cell and its uses and offers instructions for using it in the short course 'Playing with silicon cells' (Westerhoff, n.d.).[114] Moreover, information on the organisation of the initiative can be found. The next sections will focus on this organisational part and show how the organisation of the initiative started to exist as the Amsterdam Silicon Cell programme, seeking to expand internationally from there.

The Amsterdam Silicon Cell programme

The Amsterdam Silicon Cell programme outlines the research programme that underlies the Silicon Cell website. The programme is a collaboration between the BioCentrum Amsterdam (a partnership between the Swammerdam Institute of Life Sciences (SILS) of the University of Amsterdam and the Institute for Molecular Biological Sciences (IMBS) of the Free University), the Center for mathematics and Informatics (CWI) and the Institute for Informatics of the University of Amsterdam.[115] More precisely, the institutes that make up the Silicon Cell Consortium are represented by, respectively, Roel van Driel, Hans Westerhoff, Jan Verwer and Peter Sloot.

The Silicon Cell programme consists of a calculation and an experimental part. [116] The calculation part has as a long-term goal "the computation of Life at the cellular level on the basis of the complete genomic, transcriptomic, proteomic, metabolomic and cell-physiomic information that will become available in the forthcoming years". In other words, this part wants to build the actual Silicon Cell. However, as the completion of this ambition is expected to take more then a decade the work will first concentrate on three major challenges: dealing with the systematic handling of the relevant data and results, networks, space and time.

In addition, the experimental part of the research aims to add to the understanding of processes in the cell, more specifically the function of genes from which the function is not clear yet: "The availability of complete genomes has identified many genes of which the function is unknown, uncertain, or unproven. In many cases this is because the phenotype of these genes is absent, weak, or indirect; we call these the (silent and) whispering genes. Much of

---

[114] Retrieved April 30, 2007 from http://www.bio.vu.nl/hwconf/teaching/Mathbiochemie/-playsic8.htm

[115] Retrieved April 30, 2007, from http://www.bio.vu.nl/hwconf/Silicon.

[116] See for a more detailed elaboration of this objectives 'The calculation part'. Retrieved April 30, 2007 from http://www.bio.vu.nl/hwconf/Silicon.

ultimate function resides at the flux and metabolite concentration ('Metabolome') level". The experimental part inspects the functioning of large numbers of these genes systematically at the level of metabolism within the context of various small running projects that concentrate on the understanding of single-cell organisms yeast and *E. coli*.

The Amsterdam programme outlines the basic research plans of the Silicon Cell programme. Although the actual research in the Amsterdam programme deals with small organisms and researchers are working on different, relatively small grants, further research plans want to expand research efforts. The Silicon Cell initiative is staged explicitly as research with a big science ambition in programme documents, research proposals and in the stories of scientists involved. As a result, a gap between research practice and research plan emerges: while the actual science is still small-scale, the ambitions of research plans become big.

Towards an international research programme

The aspiration of bigness was first outlined in the programme entitled '*SiC!* A Dutch initiative for an international *Silicon Cells* program' (2003), written by J.L. Snoep, R. van Driel and H.V. Westerhoff.[117] The subtitle 'towards a Dutch international Silicon Cells initiative' summarizes the aim of the piece:

> This memorandum aims to boost the Dutch momentum behind what should become an international program in an area of molecular system biology that is entitled Silicon Cells (*SiC*). This initiative should result in a major proposal in the framework 6 program of the European Union, in the European Science Foundation, and in the Human Frontier and Science Program. The wider ambition is to set up an international program in which the activities in Europe, the United States, Japan and South Africa are harmonized. (Snoep et al., 2003).

By writing the memorandum, the Amsterdam consortium formed the fundament of further plans for international expansion.

In line with common arguments for big science, the scientists legitimate the enlargement of their research with techno-scientific developments. After making inventories of the different parts of the cell, the time has come to integrate all this information which is now technologically possible as well. Accordingly, the Silicon Cell Initiative is presented as a multi-disciplinary

---

[117] Retrieved April 30, 2007 from www.systemsbiology.net/sbnl/TSB.htm The programme was first written in 2002 and updated in 2003.

collaboration that integrates classical approaches in the biomolecular sciences with bioinformatics, genomics, proteomics, and metabolomics. While the Human Genome Project was only concerned with one part of the cell, the Silicon Cell involves the integration of information on all parts of a cell, which asks for a bigger scientific effort. The integration of all the available information on a cell into one model simply cannot be done in one lab or in one country. Consequently, international collaboration and collaboration with industry is required to acquire the critical mass that is able to create a Silicon Cell of a more complex organism within a reasonable timeframe.

Moreover, central coordination of research on the Silicon Cell is needed because standardization is necessary to develop a functioning model. "Currently, lots of people are working on models but this is organized on a national level. In the end, this will bring various models – a model here and a model there – but these models will not fit together and that's a big shame" (interview Westerhoff, 2005). For instance, for the combination of different models temperature is a crucial factor:

> When you have a piece of a model that is done for an organism by 25 degrees and another piece that is done at 37 degrees then they will not fit. Although the temperature maybe does not matter for the structure it just will not work when you put the two pieces of model together, which is frustrating. When they both would have been done at 27 degrees, you would not have a problem at all. (idem)

So in order to be able to put the work on models in different countries together in the end, first an agreement has to be made on temperature, as well as on other standards that are necessary to make the model compatible.

In the Dutch initiative for an international programme, the internationalisation of the Silicon Cell initiative is also legitimised in a national and European policy context. First of all, the Netherlands are positioned as a good base for the initiative. Although Germany is also an important actor in the field – the German Ministry for Education and Science has initiated a € 50 million Euro programme to make a liver cell in silico – the Netherlands should play a leading role as the Silicon Cell initiative also relates to existing Dutch research projects. The Netherlands combines a good tradition in cell physiology and other relevant scientific expertise with 'the tendency not only to compete but also to synergize'. "Dutch groups play a leading role in this field (…) why should the Dutch not play a major role in Europe?" (Snoep et al., 2003). In addition, the need for European collaboration is assessed by an overview of increasing international activity in the field. The paragraph entitled 'Why we should act quickly; rapid international developments' sketches initiatives in the United

States and Japan that Europe needs to compete with. These initiatives are put forward as reasons for joining forces on a European level, aiming at networking activities within the European Framework Programme.

Finally, the initiative is legitimised economically by addressing the relation with industry. The pharmaceutical and food industry are interested in research into the working of cells in general and especially in the working of certain specific cells, like the liver cell or yeast cell. Research into the yeast cell is interesting for companies as DSM and Unilever as the cell plays an important role in industrial processes like brewing and baking. The liver cell finds use in the medical realm, calculating the effects and conversion of drugs in the liver as well as liver regeneration after hepatitis and alcohol abuse. It is even argued that the medical world will be transformed by the shift from a largely empirical to a calculation-based operation. By outlining the applications of the Silicon Cell, it is argued that the initiative will have an economic impact: "It should be clear that the economic importance of the SiC initiative is substantial both for the near and the more distant future" (Snoep et al., 2003).

In conclusion, two ways of legitimising the bigness of the Silicon Cell initiative have become visible. On the one hand a scientific line of reasoning can be observed, taking recent scientific and technological developments as a basis for the positioning of the Silicon Cell as reasonable objective when forces are joint and coordinated. On the other hand the need for a European research network is outlined in a policy context: existing national expertise has to compete internationally via European collaboration to deliver economic benefits.

Going international with the Yeast Silicon Cell

The international ambition of the Silicon Cell initiative becomes more concrete in the plans for a European Integrated Project for the development of a Yeast Silicon Cell (Alberghina & Westerhoff, 2002). Based on earlier European collaboration on the Yeast Cell an international coordinating team of six people explored the possibilities for European collaboration in the form of an expression of interest for FP6. "We here propose a concerted European effort towards the first complete replica of one organism".[118] This resulted in an 'Information Memorandum for an Expression of Interest' for an Integrated Project called 'The Yeast *Silicon* Cell: a molecular systems biology approach' coordinated by professor Lilia Alberghina of the Department of Biotechnology

---

[118] 'YSIC: The Silicon Yeast'. Retrieved May 2, 2007 from http://www.siliconcell.net/ysic/-ysicsumm.html

and Bioscience at University of Milan-Bicocca in Italy and by Westerhoff representing the BioCentrum Amsterdam. Four other universities participated in the preparations: Göteborg University in Sweden, Catholic University Leuven in Belgium, Oxford Brookes University in the United Kingdom and the University of Stuttgart in Germany.

The aims of the collaboration have now crystallized in the development of a model cell and its application in industrial research processes (Alberghina & Westerhoff, 2002).

The goal of this proposal for an **Integrated Project** is to concentrate experimental and computational efforts toward the development of a *Yeast Silicon Cell* (YSiC) i.e. the first computer replica of an eukaryotic model organism, the unicellular budding yeast *Saccharomyces cerevisiae*. The deliverables will be progressively refined models, suitable to be tailored to specific research and industrial needs. The achievement of this goal will be of tremendous importance for:

- understanding the logic of regulation of important cellular functions (cell proliferation, apoptosis, senescence) of great relevance for human health;

- developing and testing the experimental and computational methodologies required to construct mammalian and human silicon cells;

- improving the focus of wet-lab experiments and/or reducing the need for such more expensive experiments and animal testing;

- cheap pre-testing of new drugs and new drug methodologies

- developing *silicon* cells of interest for the fermentation and pharmaceutical industries to accelerate the process of identifying and producing new and more efficacious drugs, chemical and agricultural products as well as to enable biological discovery.

Moreover, these aims are now translated into concrete deliverables between three to five years for specific users, including academic research institutions as well as industrial research and development departments, notably from high-tech SMEs.

Finally, the goal of the Expression of Interest is not only to put the Yeast Silicon Cell forward as a subject for a European scientific collaboration, but also the gathering of research partners:

> The making of an in silico replica of an entire cell, or even a significant part of it, is impossible for a single scientific group, as the diversity of the expertise required is much too broad. The aim of this project is thus to gather the best European groups working on several key aspects of yeast cell physiology in order to develop *in silico* replicas of recognizable parts of the yeast cell (modules) and then to connect these to each other. (Alberghina & Westerhoff, 2002)

They estimate that at one time about thirty to forty groups may contribute to the Integrated Project. In the process of expanding the Silicon Cell initiative, existing relations are mobilised and national and international collaborators are gathered.

With this expression of interest the vision of the Silicon Cell turns into concrete plans for research, carefully legitimised towards government and industry. Nevertheless, the expression of interest did not develop into a real European project. They decided that the work that has to be put into the writing of a concrete project proposal combined with the small chance for actually receiving funding did not make the investment worthwhile, also because Europe did not focus on systems biology yet. However, the expression of interest developed into a so-called coordination action: the Yeast Systems Biology Network.[119] A coordination action does not fund concrete research but covers coordination costs to support and promote collaborative efforts that can lead to research funding in the future.

Staging the Silicon Cell as big science

While in the first act a movement from research practice towards the articulation of the idea of the Silicon Cell could be observed, the second act stages the Silicon cell as big science. The Silicon Cell is presented as a unique and innovative concept to a wider public, mainly consisting of other scientists, policymakers and industry. Although the research is still small-scale and only first results can be shown, the research plans put the Silicon Cell forward as big science and its relevance is outlined in a science policy context as well as an industrial

---

[119] The Yeast Systems Biology Network (YSBN), a Coordination Action on Yeast Systems Biology. Retrieved June 21, 2007 from http://www.ysbn.eu

context. As a result, it seems, a gap has emerged between the actual small-scale research and the presentation of the Silicon Cell as big science. There is now a division between the actual collaboration and the presentation of the collaboration; between the research plans and the performance of those plans. As these plans did not result in funding for an actual European or international research project on the Silicon Cell, Van Driel and Westerhoff started to follow their pursuit on another level: promoting the broader research trend towards systems biology in which the Silicon Cell initiative can be embedded.

The turn towards systems biology

In 2005 Van Driel says: 'forget the silicon cell, now it is systems biology' (interview Van Driel, 2005).[120] From 2000 onwards the term 'systems biology' has become very popular in biology and can be seen as the new trend after genomics (Fujimura, 2005; Fox-Keller, 2005; Webster, 2005). In March 2002 *Science* dedicated a special section to systems biology and *Nature Biotechnology* followed with a special issue in 2004.[121] The development towards systems biology is often conceptualised as a turn from a reductionist to a holistic approach in biology (Chong & Ray, 2002). While genomics and other -omics research primarily focuses on acquiring information about the different parts, systems biology aims to make sense out of the huge amount of data through integration of information in models.

> Systems biology is about the iterative cycle of integrating data from a database into a model. Subsequently, the model can be used to make a hypothesis, which can then be tested in an experiment. The results of this experiment can then be fed back into the model again, which creates an iterative cycle that makes the model better and better. (interview Rietveld, 2006)[122]

Nowadays, systems biology has all the characteristics of a discipline: its own champions, its own journals and a yearly international conference on systems biology is organised. After systems biology champion Leroy Hood established

---

120 Prof Dr Roel van Driel is biochemistry professor leading the group Structural and Functional Organization of the Cell Nucleus at the Swammerdam Institute for Life Sciences of the Faculty of Science of the University of Amsterdam. In addition, he has been appointed Faculty Professor at the Faculty of Science.
121 Science, volume 295, 1 March 2002; Nature Biotechnology, volume 22, October 2004.
122 Dr. Luc Rietveld is working as a policy officer for ZonMW and the Netherlands Genomics Initiative and dedicates part of his time to coordinating systems biology research.

the first Institute for Systems Biology in Seattle, similar institutes are now appearing all over the world. In addition, systems biology is often seen as a way to acquire research funding: "Some people view systems biology as another complot of the molecule-mafia to acquire money" (interview Rietveld, 2006). This is underlined by an article in *The Scientist* (Wiley, 2006) that invites its readers to follow the hype by describing 'Five simple steps towards a systems biology approach'.[123] Moreover, systems biology is presented as a scientific enterprise that will deliver profits (Mack, 2004). It will influence all fields where live organisms or processes can play a role, varying from health, agriculture and food to industrial processes, energy and the environment.[124]

This section discusses the turn towards systems biology. As the Silicon Cell and other modelling efforts can be seen as part of this broader development, Westerhoff and Van Driel started to proliferate their work as systems biology. They turned their attention to the promotion of this new approach within academia and towards policy on a Dutch and European level. This section follows the two scientists on their systems biology adventures. By subsequently analysing their performance in and outside academia, I will show how the new organisation of science is interwoven with the establishment of a new scientific approach. Building on their experience with the Silicon Cell initiative, Westerhoff and Van Driel contribute to the organisation of the new field of systems biology.

Performing systems biology inside academia

To encourage the new development towards systems biology that emerged at the beginning of the new millenium, Westerhoff and Van Driel engaged in legitimising and organising systems biology within a Dutch and European scientific context. Knowing that 'systems biology' is a contested concept, they started conversations with other scientists about new developments: "We were talking about systems biology or integrative biology. There are all kind of words for this – buzzwords – that are used and misused in various context. So when you ask three people about the meaning of systems biology, you get three different answers" (interview Van Driel, 2005). Nevertheless, the Silicon Cell scientists view the developments that are indicated with the term systems biology as a fundamental step in the life sciences:

---

[123] Retrieved June 6, 2008 from www.the-scientist.com/article/display/23585/

[124] Remacle & Benediktsson (2006) 'Proposal workshop on systems biology'. Retrieved June 8, 2008 from ec.europa.eu/research/biotechnology/ec-us/docs/remacle_20_july_12-45_en.pdf

We – and I refer to a few European and international researchers, and the Silicon Cell is part of that – are gradually able to understand cells, organisms and also multi-cellular organisms in such a way that we can translate the knowledge into molecular models. So on a molecular level you make models in your computer on how molecules in time and space interact with each other (…) With this, we ultimately can understand how life works on a molecular level. (idem)

In an attempt to promote this exciting development and share ideas with fellow scientists, Westerhoff and Van Driel started to talk to various prominent Dutch scientists: "We just made an appointment and asked how they thought about current developments" (idem).

Unintentionally, these conversations stirred up a harsh debate on the merits of systems biology in the Dutch biological community. After they talked with Professor Plasterk, who is known for his work on developmental biology and zebra fish, he wrote a punchy article about systems biology and the development of a silicon cell. In the *Bionieuws* magazine of the Netherlands Institute for Biology he reacted against the new hype of systems biology, calling it a danger, an illness and the emperor's clothes (Plasterk, 2003). His main argument: systems biology creates a false opposition between understanding the whole and its parts, as the understanding of the whole depends on the detailed understanding of the parts. Consequently, systems biology will not lead to new knowledge as it is not a new approach in biology. In response, Westerhoff and Van Driel wrote an article in which they present systems biology as an approach that delivers "fascinating results" as the interaction between parts can lead to additional properties that are fundamental for the biological function (Van Driel & Westerhoff, 2003).

The two scientists underlined their belief in systems biology with words and actions. Especially Westerhoff has been involved in defining what systems biology entails. In a presentation, he defines systems biology as 'nothing vague, but highly exciting'. Overviewing various definitions of systems biology, he concludes that they highly converge, leading to the following overall definition on a slide of Hans Westerhoff.[125]

What is System Biology, then?

- The study of the emergence of functional properties that are present in a biological system but not in its individual components

---

[125] Retrieved June 15, 2007, from http://www.systembiology.net/whatsb_files/v3_document.

- From molecules to organism/ecosystem (and back)
- Quantitative (modellable) experiments
- Models relating to the only experimental reality
- Not just analysis, also understanding

However, this is just the start of a larger definition exercise. Several articles on systems biology have been written, such as a publication in *Nature* in 2004 entitled 'The evolution of molecular biology into systems biology' that Westerhoff wrote together with Bernard Palsson, professor of Bioingineering at the University of California in San Diego (UCSD). In addition, Westerhoff edited a handbook on systems biology together with Lilia Alberghina, already known from the European project on the Yeast Silicon Cell.

Systems Biology; definitions and perspectives (Alberghina & Westerhoff eds., 2005) & Systems Biology; philosophical foundations (Boogerd et al., 2007).[126]

In the pursuit to define what systems biology actually entails also a philosophical perspective on systems biology has been explored. Together with colleagues from Amsterdam and Stellenbosch, Westerhoff organised a symposium on the philosophy of systems biology in 2005.[127] The symposium tried to establish

---

[126] Coverimages courtesy of Springer and Elsevier.
[127] Retrieved June 21, 2007 from http://www.systembiology.net/philosophy

philosophical foundations for systems biology, by relating different views on systems biology to various philosophical traditions in biology. After looking into the methodology of systems biology and the role of theory, models and simulation in biology, the different ideas on organization in biology and their implications for research practice were explored. Ultimately, the philosophical foundations of systems biology were presented as fundamentally different from those of molecular biology. The symposium resulted in an edited volume in which entails the different reflections on the philosophy of systems biology (Boogerd et al., 2007).

Next to this definition work, various actions underlined the importance of systems biology. First of all, a new website was built, located at http://systems-biology.net. Now this website is the umbrella that spans national and European activities in systems biology, including the silicon cell website. In addition, various workshops and conferences have been organised on a Dutch and international level. For instance, Westerhoff has been involved in the organisation of the fourth International Conference on Systems Biology in Heidelberg in 2004. Moreover, educational activities were organised. The Free University of Amsterdam started an educational program in systems biology.[128] From 2004 summerschools on systems biology have been organised in the Netherlands and from 2005 biannual SysBio courses in Gosau, Austria in cooperation with the Federation of European Biochemical Societies (FEBS).[129] The various articles, books, courses, workshops and conferences, which come together on the new website, have certainly put systems biology on the scientific agenda.

Performing systems biology outside academia

If scientists had to be convinced of the new approach, systems biology also needed to earn a place for itself on the science policy agenda. Westerhoff and Van Driel contributed to the development of policy for systems biology on a Dutch and European level by establishing System Biology NetherLands (SBNL) and a European Systems Biology Initiative. System Biology Nether-Lands was formed on the basis of a meeting in November 2002 organised by Van Driel, Westerhoff and others in cooperation with the division Earth and Life Sciences (ALW) of the Dutch Research Council (NWO).[130] As the strategy of ALW-NWO focused on the development of partnership, the scientists

---

[128]Retrieved June 21, 2007 from http://www.systembiology.net/topmaster/

[129] Retrieved June 21, 2007 from http://home.deds.nl/~rogier/sb2004/; http://www.issb.org/; http://www.univie.ac.at/sysbio2007/

[130] Retrieved June 21, 2007 from http://www.systembiology.net/sbnl/index.html

explored the possibilities of national and international collaboration in the context of systems biology. Goal of the workshop was the development of a national research agenda for systems biology to contribute to European programmes. The group of scientists from different Dutch universities chaired by Van Driel, had several follow-up meetings and later also included policymakers and industry.

The Dutch Industrial Platform on Systems Biology also incorporated actors from science policy – from the Dutch research council (NWO), the Netherlands health research council (ZonMW) and the Netherlands Genomics Initiative (NGI) – and research managers from Dutch companies DSM, Unilever and Organon. The platform explored cooperation between academia, government and industry, resulting in the memorandum 'SysBioNL; a Dutch programme for food and pharma' (Van Driel et al., 2004). This programme explicitly focused on the application of systems biology knowledge in the pharmaceutical, food and biotech-industry. They prepared a proposal for the so-called smart-mix, which was declined.[131] Nevertheless, the initiative put systems biology on the agenda of Dutch research policy by assembling central actors and stimulating attention for systems biology in science policy organisations.

Going beyond the national border, Van Driel and Westerhoff have been involved in the organisation of systems biology policies on a European level. Next to influencing the policy of the Directorate-General for Research of the European Commission towards systems biology, they sought alternative ways to promote systems biology within Europe which they found in the context of the European Science Foundation, the European federation of national research councils.[132] They chaired a commission that performed a 'Forward Look on Systems Biology', an exploration of possibilities for European research on systems biology. Again, they talked with lots of people – but now on a European level – and organised several small workshops to which they invited people from academia, industry and European science policy. "These were very effective and in the end we had a two days conference in Gosau, Austria with some international hot-shots attending" (interview Van Driel, 2005). The conference formed the basis of a report, entitled *Systems Biology: a Grand Challenge for Europe* (ESF, 2005).

The report presented 'A European Action Plan' for the creation of a European Systems Biology programme. The plan consists of seven actions

---

[131] Smart-mix is a Dutch funding initiative for public-private collaboration. Retrieved June 7, 2008 from http://www.smartmix.nl/

[132] Retrieved June 21, 2007 from http://www.systembiology.net/esfflshort.pdf

geared to building an infrastructure for systems biology research in Europe: the set-up of a task force to develop a European road map; establishing a consortium of European Reference Laboratories; cooperation between industry, academia and charities; public acceptance; training and education; financing; and the establishment of a European Systems Biology Office (ESBO). Especially the financing of this whole systems biology enterprise is seen as an important challenge:

> The Europe-wide approach proposed here will require a higher level of funding than that provided by vehicles available now. We propose that a new financial model is developed based on cooperation between national, international, industrial, charities and EC-related organisations. It will be a major challenge to align this heterogeneous set of parties so as to produce a synergistic cooperative programme. (ESF, 2005: 6)

In other words, as the European Systems Biology programme is of a different scale than current scientific endeavours in biology, a new order has to be created within the European scientific world.

Although this new order is not a reality yet, the scientists are still working on making it happen. They are now part of the ESF Task Force on Systems Biology that is trying to pursue recommendations of the ESF Forward Look, establishing ESBO as well as a Grand Action on Systems Biology (GRASB).

> This programme should consist of a portfolio of coordinated activities aimed at an integral activity entitled 'Networks for Life'. The portfolio of activities should build on the major Systems Biology activities that the taskforce members have put in place already, and constitute the world's largest and most effective single Systems Biology programme. (Taskforce on Systems Biology, 2007: 5)

In addition, Van Driel and Westerhoff are involved in the design of a European Research Era for Systems Biology. ERASysBio aims to create a systems biology policy that integrates national systems biology initiatives: "The countries involved in ERASysBio are fully aware of the need to join forces in creating international policy on systems biology, to enable competitiveness in the field" (ERASysBio Partners, 2007: 4). In sum, Van Driel and Westerhoff have been continuously working on the construction of a large-scale collaboration in systems biology.

An article entitled *Fundamental issues in systems biology* (O'Malley & Dupre, 2005) that primarily reflects on epistemological issues within systems biology, indicates at the end towards the importance of research into the social factors that shape systems biology:

> All these questions barely touch upon the fact that the future of systems biology will be shaped as much by social factors as by scientific and philosophical ones, with different ways of thinking about systems evolving within a context of 'big biology' funding, industry expectations, and conflicts between diverse disciplinary cultures provoked by the interdisciplinary mandates inherent in systems biology. (idem: 1274)[133]

However, the story of the Silicon Cell and its embedding in systems biology shows that ways of thinking do not just evolve within a certain static context, but that this context evolves in interaction with the development of new ways of thinking. The performance of Westerhoff and Van Driel inside and outside academia illustrates how their contribution towards the establishment of a new approach within biology cannot be separated from their efforts to create a new scientific order. In other words, the scientists are involved in what Goodman (1978) has called 'worldmaking'.[134] In this case the set-up of a scientific collaboration can be seen as an attempt to create a new world of biology.

Review

In this chapter I analysed the construction of the Silicon Cell initiative and later efforts in systems biology in three acts. First, visions of the cell in a computer emerged, followed by the staging of the Silicon Cell as big science and a turn

---

[133] The authors are based at the ESRC Centre for Genomics in Society (Egenis) based at the University of Exeter in the United Kingdom, which studies systems biology as a new direction in genomics. Retrieved June 21, 2007, from http://www.centres.ex.ac.uk/egenis.

[134] "Much but by no means all worldmaking consists of taking apart and putting together, often conjointly: on the one hand, of dividing wholes into parts and partitioning kinds into subspecies, analyzing complexes into component features, drawing distinctions; on the other hand, of composing wholes and kinds out of parts and members and subclasses, combining features into complexes, and making connections. Such composition or decomposition is normally effected or assisted or consolidated by the application of labels: names, predicates, gestures, pictures, etc. Thus, for example, temporally diverse events are brought together under a proper name or identified as making up 'an object' or 'a person'" (Goodman, 1978: Ways of Worldmaking).

towards systems biology. After the play it is time for a review. What does the construction of the Silicon Cell initiative tell about the building of scientific collaboration? I will show the interplay between backstage and front stage in the building of collaboration and indicate how the building of collaboration influences the role of scientists involved. Finally, I reflect on the difficulties of building collaboration in systems biology.

Building collaboration

When analysing the construction of scientific collaboration in detail, it becomes apparent that the subject and organisation of scientific collaboration evolve together. The first acts demonstrate the close relation between science and its organisation by showing how the Silicon Cell is gradually shaped, simultaneously as scientific object and as collaboration. The Silicon Cell has different meanings during the different stages. In the first act, the cell comes into being as a means to solve a dispute and a method to perform research. However, it gradually becomes a mechanism, a general principle that makes a transition from wet to dry biology possible. The Silicon Cell is simultaneously a vision and a concrete model of (part of a) cell in a database. In the second act, the idea of the Silicon Cell is turned into a subject of large-scale collaboration and it acquires various applications. The cell becomes a means to improve human health – for instance curing cancer – and it also turns into an industrial tool. In the third phase, the cell becomes part of the broader development towards systems biology. In sum, the meaning of the Silicon Cell changes continuously throughout the history of the collaboration.

The different definitions of the Silicon Cell are related to different organisational activities. In the first act, the idea of the Silicon Cell emerges out of research practice in a traditional academic environment; with PhD students performing their research and collaborating within a research group headed by a professor. However, when the research results are presented to various audiences – including science, policy and business – the idea of the Silicon Cell becomes articulated and staged as big science. As subject of large-scale collaboration, the network around the Silicon Cell starts to grow and collaboration expands the boundaries of the research group. First, expansion takes place along existing relationships and scientists from other groups, departments and institutes become involved. Later on, new relationships are established and collaboration with policymakers and industry takes shape. In this context also a transition towards systems biology takes place. Instead of building the Silicon Cell collaboration from the bottom-up, now it is built top-down by embedding it into a larger international organisational effort. In this way, the organisation of the research changes along with the idea and meaning of the Silicon Cell, moving from small-scale research to a large-scale international network.

In addition, analysing the construction of collaboration as a performance shows how scientific collaboration is shaped through interaction with various publics. Although the origin of the Silicon Cell can be found in actual research that takes place behind the closed doors of the laboratory, or backstage, the articulation of the idea of a large-scale collaboration materialises when presenting the research to various audiences front stage. The idea of the Silicon Cell emerges in a debate with other scientists and first developed when Westerhoff had to present the research results to the scientific audience. Later, the website and the various proposals were designed to sell the cell, also transgressing the borders of science by addressing policy, industry and basically everybody surfing the World Wide Web. So it is in the presentation of the research to a broad audience that the Silicon Cell gradually became defined as a large-scale collaboration. In other words, the staging of collaboration takes place in crossing the borderline between backstage and frontstage, in the translation of research, its meaning and potential towards a larger audience.

As a result, the distinction between backstage and front stage makes visible how scientific collaboration does not only mean performing research together, but also involves the presentation of collaboration. The case of the Silicon Cell illustrates how during the construction of collaboration backstage and front stage gradually begin to diverge. The first act shows small-scale science backstage that is also presented as small-scale science front stage. But then the performance starts to change. The second and third act demonstrate how the Silicon Cell and systems biology are carefully staged as big science through the creation of plans with visions into a new world of biology. In contrast, the actual research practice keeps having a small-scale character as the plans did not materialise. So while the Silicon Cell becomes big science front stage, it stays small backstage. This process of presenting bigness is crucial in building collaboration, as research proposals are required to acquire funding for actual collaborative research. However, ideally the plans will be funded and the science will become big as well, rendering front stage and backstage equally big.

Changing roles of scientists

Analysing the process of building collaboration from a dramaturgical perspective also shows how the role of scientists transforms. The creation of collaboration in systems biology requires that the scientists involved play different roles. They change from quite normal professors running their own lab into actors in a new research production. While they were already used to play their role as head of a research group presenting to fellow scientists and funding organisations, Westerhoff and Van Driel now have to perform collaboration in front of different audiences, varying from scientists to policymakers and industrial parties. Put differently, the construction of collaboration requires them to change hats and fulfil the roles of network manager, lobbyist, policymaker and negotiator. These new roles require new knowledge, new skills and new scripts. Apparently, the creation of a new order in biology – making biology big – goes beyond the traditional scientist's role and calls for the performance of multiple roles.

During the process of constructing collaboration, the roles of scientists gradually started to shift. In the first act, Westerhoff performs the role of head of the laboratory supervising PhD research. After some successful experiments, research results are presented to fellow scientists and the idea of the Silicon Cell emerges. When in the second act the Silicon Cell is presented as big science, roles start to change. Presenting the Silicon Cell to a wider public, Westerhoff teams-up with the Stellenbosch group and becomes a website builder. In addition, together with Van Driel and other scientists involved, the Amsterdam Silicon Cell programme is written and presented to fellow scientists and policymakers. In subsequent efforts, the Silicon Cell is staged as big science, which requires the scientists to gather and manage an international network of scientists and lobby with funding organisations and policymakers. In the third act, the switch to systems biology turns Van Driel and Westerhoff into policymakers and negotiators. In putting systems biology on the Dutch and European agenda, they start to lobby inside and outside academia. During the making of plans, they discuss and negotiate with fellow scientists, policymakers and businessman. Finally, with the Forward Look they perform the role of policymakers, constructing European policy for systems biology. Significantly, from being professors heading a laboratory Westerhoff and Van Driel have gradually turned into actors who perform multiple roles.

Although this role change was necessary and the scientists enthusiastically embraced their new roles, the multiple roles also caused some serious tensions in terms of time, practice and values. First of all, both scientists feel they have not enough time. "We are extremely busy, we are spending time with NWO, in Germany and Europe. On top of this I also have an appointment of 60% in Manchester. It simply makes me crazy" (interview Westerhoff, 2005). Roughly, Westerhoff spends about 60% of his time on lobbying and policymaking, 30% of his time on management and 10% of his time doing science. Also Van Driel told me that his activities for systems biology took some 50% of his time (interview Van Driel, 2005). He did not mind it, though: "as long as you enjoy it" (idem). However, Westerhoff identifies a tension between being a scientists and performing other roles:

> This simply cannot be combined with science. When you are always on committees, there is a danger that you don't know what is going on in the lab anymore. Although things are still going okay now, this is a present danger. When for instance you as a professor suggest an experiment and your group tells you: 'That's impossible because this or that. Didn't you know?' and you in fact did not know, you do have a serious problem. Because in the end they will not listen to you anymore. (interview Westerhoff, 2005)

Clearly, it is very difficult to combine the role of leading scientist with other roles involved in building collaboration.

Next to time, the issue of experience turns out to be a bottleneck. Van Driel and Westerhoff realise that their role changes but express that they do not have the experience to perform their new characters in a good way and carry on towards the end. "We run into all kind of non-scientific issues we don't know about" (interview Van Driel, 2005). They present themselves as amateurs that lack experience to perform the new roles properly: "I should have realised earlier how politics works (…) I am not good at this and I have not been educated for this" (interview Westerhoff, 2005). Moreover, this lack of experience prevents the scientists from carrying their efforts forward. Although their activities do find resonance, it has thus far mainly led to written plans, authorized by partners from academia, government and industry. Turning the plans into concrete action will require further steps, but the scientists are not sure what role they have to play and what script is appropriate to actually set-up the collaboration: "We don't know how to go on. At this moment all advice is welcome" (interview Van Driel, 2005). The scientists expect that turning words into deeds will make negotiations much tougher, especially when it concerns money. As a result, they do not feel that they are able to take the process a step further and would rather see someone else take over from here so they can return to the lab again. While the scientists enthusiastically perform their new roles in the first acts, they have no script for the final act of setting up collaboration.

The lack of experience with lobbying for large-scale research is a more general problem within the life sciences. European policy officer Dr Henriette van Eijl acknowledges the problem: "Biologists often think too small" (interview Van Eijl, 2005). This can be contrasted to such other scientific disciplines as physics and space research that are used to think big and are very well organised for large-scale research. In contrast to biology, these two fields do not have a very clear societal relevance and still they are able to get funded. As a result,

Van Driel would really like to know how CERN physicists organise their lobbying and he also wonders about space research: "NASA in the US seems to have a great PR. Even if spacecrafts crash, they are still be able to continue their research and get lots of money. So we can learn something from them" (interview Van Driel, 2005). Nevertheless, lobbying in biology slowly starts to take shape in interaction with the construction of large-scale collaboration: "fifty percent deals with content, the rest has to be learned" (idem).

The different roles also bring along different values that can collide and cause tensions. In line with Mertonian values Westerhoff continues to stress his content-related view and his objective stance: "I have the idea that the success we have is only based on the powerful argument; I only tell what there is and do not play politics" (interview Westerhoff, 2005). In addition, when asking him about strategies he firmly denies using any: "Strategy ... tactics ... I do not engage in it because it is antithetic to the academic endeavour in which you search for truth, try to be objective and collaborate with your colleagues" (idem). However, Westerhoff also reveals several good moves that Van Driel and he made in the effort to set up collaboration, which could well be called strategic, and also their actions clearly show strategic insight. Nevertheless, they have an ambivalent stance towards performing their new roles. On the one hand they enjoy it and see it as a necessity, but on the other hand they clearly prefer their role as traditional scientists: "I do not have the inspiration to end my scientific career to become organiser of systems biology" (interview Westerhoff, 2005). As a result, the scientists perform multiple roles with multiple scripts, but keep presenting themselves as traditional scientists.

Cell death?

While the Dutch scientists have been working to put a collaboration in place since 1997, the Silicon Cell has not become the international project that was envisioned. Only the future will tell if Europe will see the world's largest collaboration in systems biology and if the Silicon Cell will grow. But what has made realisation of this research project so hard? The scientists involved have various explanations. Next to evaluating their own performance – concluding that they neither have the skills nor the time to build collaboration – the scientists focus on the shortcomings of science policy. First of all, they claim that the Dutch national science policy context causes problems when building collaboration around the Silicon Cell. Although the Netherlands has a tradition in cell physiology that would serve as a scientific basis for collaboration, the political context complicates the story. "The Netherlands are a small country and things start moving very slowly" (interview Van Driel, 2005). Especially when comparing the Netherlands to the United States it becomes visible that science is not high on the Dutch policy agenda and that new developments are relatively slowly incorporated in policy.[135] Nevertheless, the scientists acknowledge that the establishment of the Netherlands Genomics Initiative (NGI) to advance genomics research has improved investments in biology research.[136] However, funding for the biosciences is not concentrated in one single research council, but divided between the new NGI that was established in 2003, the traditional life sciences department of the general research council (NWO-ALW) and the medical research council ZonMW. As they all have their own internal politics, this makes acquiring funding or lobbying for large-scale collaboration quite complicated from a scientific perspective.

In addition, scientists experience that Dutch science policy has a strong emphasis on application, which asks for interaction with industry: "Everything needs to be applied, while in the US – a capitalist country pur sang – research at the NIH is very fundamental" (interview Westerhoff, 2005). As a result, scientists need to engage with industry to put systems biology on the agenda. Although industry is becoming very interested, setting up scientific collaboration with the industry in systems biology comes with its own difficulties: "They do not have to be convinced anymore, but it is now about how to structure the research and this is what makes things complicated" (interview Van Driel, 2005). First of all, industry is not used to research on this scale and the coordination of research in different large companies in the Netherlands is quite a new experience. In addition, the long-term character of systems biology research is a problem for industry in terms of cost-efficacy. Moreover, it is difficult for scientists to gain insight into the agenda of industry: "Each industry has its own agenda and different industries have different agenda's and they only show you part of it" (idem). As a result, collaboration with industry also complicates the construction of large-scale collaboration.

When going beyond the Dutch national context the establishment of international collaboration turns out to be very difficult. "In biology, science policy is not geared towards large-scale collaboration" (interview Van Driel,

---

[135] While the budget of the National Institutes of Health in the United States has doubled to enable new developments in biology, investments in science have slowly gone down in the Netherlands. While in 1995 the Dutch government invested 0.386 % of its BBP in research, in 2006 this was 0.290 % (Retrieved June 8, 2008 from http://www.vsnu.nl/web/show/id=77895/-langid=43/framenoid=41685). According to Westerhoff, the lack of money for research has a serious impact on funding rates and research performance: "in the Netherlands we are spending much more time on writing research proposals" (interview Westerhoff, 2005).

[136] Chapter 5 will further explore science policy and the role of the Netherlands Genomics Initiative.

2005). On the one hand, countries do not want to spend their money on research which is performed in another country; this inhibits the collaboration between countries that is so crucial in the case of systems biology: "When every country has its own small projects, it's a waste of money. We have to standardise to make something useful" (interview Westerhoff, 2005). On the other hand, European research funding that explicitly wants to stimulate international collaboration brings its own problems. First of all, funding rates are very low while it takes a lot of time to prepare a proposal. In addition, the politics of distribution over various European countries plays an important role. However, especially the scale and the newness of the systems biology plans make it difficult to set up European collaboration. The EU funding structure is not geared towards new scientific developments as it is fixed in advance: "The funding structure of the EU is very inflexible because within the Framework Programmes everything is nailed for five years in a row" (interview Van Driel, 2005). Moreover, even for European standards the collaboration that Van Driel and Westerhoff have in mind is quite large and therefore requires the building of a new collaborative structure. The scientists therefore try to create this new structure through the forward look of the European Science Foundation.

When taking a more analytical perspective, it becomes clear that the building of large-scale collaboration in systems biology proves to be difficult because it also involves the creation of an environment that enables this new type of collaboration. In their overview of scientific collaboration. Shrum et al. (2007) distinguish different ways in which collaboration is established.[137] From this categorization, two types of formation can be deducted: conventional and unconventional collaboration formation. When scientists are used to collaborate and funding is relatively easily available, the setting-up of collaboration is characterised as conventional or business-as-usual. However, when collaboration is not common practice and (funding) policy is not geared towards collaboration, the formation of collaboration becomes unconventional or entrepreneurial. This is exactly what makes the building of collaboration in the case in the Silicon cell initiative and subsequent initiatives so difficult. Not only the collaboration itself has to be constructed, but also the policy environment and funding structures have to be built. It is precisely the creation of a policy and funding space for large-scale collaboration in systems biology that is experienced as an important bottleneck by scientists involved in constructing collaboration.

---

[137] For a detailed explanation of different ways in which collaboration can be formed, see Chapter 1 in Shrum et al. (2007).

Nevertheless, the scientists are making important progress by putting systems biology on the agenda. Van Driel already diagnosed the situation as follows: "Especially the last years a lot has happened. Lots of things have been put on track and some trains are running. However, they have not reached the station yet" (interview Van Driel, 2005). In the past years systems biology has certainly acquired a place in Dutch and European research policy and quite some research efforts are already taking place. In the Dutch context the Netherlands Genomics Initiative has established a Centre for Medical Systems Biology[138] and systems biology has become part of the strategies of major funding organisations like NWO (2006). In addition, the Netherlands is involved in SysMo, a European transnational funding and research initiative on the systems biology of micro-organisms. [139] On a European level, some systems biology initiatives were already funded in the sixth Framework Programme and within the new seventh Framework Programme systems biology has become a central theme.[140] More concretely, Westerhoff is for instance participating in the Yeast Systems Biology Network. However, Hans Westerhoff has decided to pursue an important part of his research outside of the Netherlands, as a professor at the new Manchester Interdisciplinary Biocentre which entails one of the six centres of excellence in systems biology established in the United Kingdom.[141] Roel van Driel is now directing the Netherlands Institute for Systems Biology, established in 2007, which stimulates collaboration in systems biology.[142] It seems, then, that these scientists have contributed to creating a window of opportunity for systems biology and collaboration after all.

---

[138] Retrieved June 21, 2007 from http://www.cmsb.nl.

[139] Retrieved June 19, 2007 from http://www.sysmo.net.

[140] Within FP6 projects activities are for instance: NucSys: A Marie Curie Training Network (Retrieved June 21, 2007 from http://www.uku.fi/nucsys); BioSim: A Network of Excellence for BioSimulation. (Retrieved June 21, 2007 from http://biosim.fysik.dtu.dk:8080/biosim); the YSBN (Retrieved June 21, 2007 from http://www.ysbn.eu).

[141] Retrieved June 21, 2007 from http://www.mib.ac.uk.

[142] Retrieved September 12, 2008 from http://www.sysbio.nl.

CHAPTER 5

Developing a new vaccine
The innovation epidemic

We cannot already make a vaccine against the next 'pandemic' because the virus continuously changes. We do not even know for sure that it will be H5. So, what in my view should happen now is to take. for example, a H5 or H9 subtype, make a prototype – a candidate vaccine that is already adjuvated – and test it with humans. However, these are very expensive studies; we are talking about a multiplicity of ten million euros. Industry will not finance this of course, because those vaccines cannot be sold as it actually is technology development and a whole new infrastructure should be developed. However, this is something that should be done right now. When starting at the moment of the outbreak of the pandemic, a year is needed before proving the safety and efficacy of the vaccination and another half a year to produce the vaccine. That means that one and a half year has passed and experience teaches us that the pandemic is already gone by then. It has already passed around. So I definitely think that now is the time to experiment with prototypes of vaccines to prove which technologies should be used to make effective and safe pandemic vaccines.[143]

Talking is Professor Ab Osterhaus, a virologist who is head of the department of virology of the Erasmus Medical Centre in Rotterdam. He is an expert on pandemics who frequently appears in the Dutch media. What is he talking about? First of all, he speaks of the risk of an influenza pandemic in the context of the H5N1 virus, commonly known as bird-flu. The sudden spread of SARS

---

[143] Video-interview with Professor Ab Osterhaus on the website of 'Erfocentrum' (the Dutch national knowledge and educational centre for heredity and medical biotechnology), 24 October 2005. Retrieved October 29, 2007 from http://www.biomedisch.nl/film/vogelgriepvirus.php.

in 2003 made scientists, governments and media increasingly pay attention to the risk of pandemics. After the Spanish flu in 1918 (over 40 million deaths estimated), the Asian flu in 1957 (40-50 % of the world population affected and about 2 million deaths) and the Hong Kong flu in 1968 (between 1 and 3 million deaths estimated) it is now believed that a next flu pandemic is likely, if not inevitable (WHO, 2005a; WHO, 2005b; Health Protection Agency, 2006; Kolata, 1999). From 2004 onwards, bird-flu has been seen as the most threatening virus with the potential to cause the next global flu pandemic. Risks are higher than ever, as within our modern society with its global transport infrastructure a virus that infects humans will spread even quicker than during previous pandemics (Bijker, 2006). This is why governments are preparing for such a global outbreak. In this respect, for example, the president of the United States has stated: "Together we will confront this emerging threat and together, as Americans, we will be prepared to protect our families, our communities, this great Nation, and our world".[144]

Within this general context of fear for a flu pandemic, Osterhaus is talking about influenza research and the development of a vaccine, while simultaneously unfolding a research agenda. Because of the certainty of a coming pandemic and the uncertainty of the exact composition of the virus that will cause human infection, he argues that we have to act now in order to be prepared. With this reasoning Osterhaus eloquently creates a sense of urgency by mobilising the future in the present (Van Lente, 1993; Brown et al., 2000). Moreover, he gives his view on who should take action. As he clearly points out that industry will not perform the necessary research, because there is nothing for them to gain yet, research should be financed by government. Osterhaus thereby touches on the complex relationship between science, government and industry in innovation. In preparation for a possible pandemic, governments have to invest in public research, which eventually can lead to the development and commercial production of vaccinations or other forms of therapy by the pharmaceutical industry. This makes research into flu an example of research in which the domains of academia, government and industry become intricately linked.

As already argued in Chapter 2, application and innovation are characteristic of contemporary big biology. In line with contemporary theories and policies of innovation that increasingly challenge the boundaries of academia by asserting that innovation occurs at the intersection of academia, government

---

[144] George. W. Bush in the preface of the 'National strategy for pandemic influenza' (2006), published by the National Security Council, Washington D.C. Retrieved October 29, 2007, from http://www.whitehouse.gov/homeland/nspi.pdf

and industry (Elzinga, 2004; Etzkowitz et al., 1998; Etzkowitz & Leydesdorff, 2000; Gibbons et al., 1994; Hessels & Van Lente, 2008; Shinn, 2002), this chapter investigates the way in which these domains become entangled in the building of the VIRGO project. VIRGO is an academic-industrial collaboration that aims to develop innovative therapies against flu and is directed by Professor Osterhaus. Where in the previous two empirical chapters the involvement of government and industry in scientific collaboration was only briefly touched upon, this chapter explicitly looks into the role of these actors in building large-scale research in the life sciences. More precisely, the chapter investigates how the connections between academia, government and industry materialise within the VIRGO project. However, I will show that the merging of these different domains is hardly a straightforward matter. Next to the intermingling of different domains and the crossing of boundaries, I also demonstrate how borders are realigned or kept in place.

To show how science policy and its emphasis on innovation shapes research collaboration, the chapter will start with a general overview of government policies in the life sciences, illustrating how they focus on the stimulation of innovation. Against this general policy background I investigate the way in which policies shape the academic-industrial VIRGO consortium. I will analyse how the research project originates from a new approach to virology research within academia and how it becomes a collaboration with industry within the context of the Dutch governmental initiative to stimulate genomics research. Subsequently, I explore how academia, government and industry become intertwined in the VIRGO project. How do the structure and practices of the research project enable or prevent the entanglement of the different domains? Finally, I will reflect on the emphasis on innovation in science policy and the way these policies reconfigure scientific collaboration.[145]

---

[145] My analysis of science and innovation policy and the VIRGO consortium is based on primary and secondary sources, comprising websites, policy documents, scientific articles and media coverage. In addition, I have performed in-depth interviews with a number of key informants in academia, government and industry: Andeweg, Van den Berg, De Geus, Dons, Horning, Hubbard, Lelyveldt, McCarthy, Musselwhite, Ogg, Sanders, Spek and Van Oort (for a complete overview of interviews, see Appendix A). Moreover, I have attended meetings related to science and innovation policy (for an overview of meetings, see Appendix B).

Life sciences policy

The first word that comes to mind when looking into life sciences policies is resemblance. The life sciences have become subject of governmental strategies all over the world, stimulating research as well as innovation. Almost every country or region wants to become part of the top when it comes to the promising new developments in genomics and post-genomics research, also benefiting from its economic value – from Silicon Valley to Kenya and from Europe to India (Cortright & Mayer, 2002; EC, 2002; Ngubane, 2001; Sibal, 2005). When browsing through the different strategy documents, always the same arguments are presented: starting with the importance of research into life for health, agriculture industry and environment (also referred to as red, green and white biotech); picturing the current state of affairs; and developing a strategy aimed at becoming one of the world's leading centres for research and innovation. Similarly, the strategies have the same basic ingredients: the establishment of research centres and networks to stimulate research and collaboration; the stimulation of commercialisation of knowledge; and the embedding of science in society by the establishment of programmes or centres that look into the ethical, legal and social aspects or implications of genomics research.

The life sciences strategies all stress the importance of innovation. In line with these strategies, policy plans to create innovation are very similar. Countries often look at successful policies of other countries in order to imitate. In Europe the United States are often seen as an example for innovation (EZ, 2002). However, also within the United States innovation takes place through imitation. Following the example of the success of Silicon Valley, regions are increasingly trying to become an innovative region. All states in the United States have developed innovation policies concerning the life sciences. From Massachusetts' 'MassBiotech 2010: Achieving Global Leadership in the Life-Sciences Economy' (MBC, 2002) to 'Biotechnology in Hawaii: A Blueprint for Growth' (PMP Public Consultancy, 1999). This even led to the development of recipes for innovation, as 'A Governor's Guide to Cluster-Based Economic Development' (National Governors Association, 2002) and 'Building State Economies by Promoting University-Industry Technology Transfer' (Tornatzky, 2000).

Policy measures to create innovation generally exist of the stimulation of academic-industrial collaboration, the building of innovative clusters and the commercialisation of knowledge. As in the case of VIRGO, the construction of academic-industrial collaboration is stimulated, for instance, in the LINK programme of the English Biotechnology and Biological Sciences Research Council in cooperation with the Department of Industry.[146] In turn, Singapore promotes academic-industrial collaboration through the creation of Biopolis "a purpose-built biomedical research hub where researchers from the public and private sectors are co-located".[147] This last example also resembles the creation of innovative clusters; places where academia, industry and government are in geographical proximity enabling collaboration. Clusters focussing on the life sciences are set up throughout the world and have such names as Medicon Valley, Cellulose Valley, Food Valley, Biotech Bay, Biotech Beach, BioCapital, BioCorridor, BioGarden, BioForest and Genetown.[148] In addition, academic research is commercialised through translation processes that include patenting, technology transfer and the creation of new companies, enabled by incubators and special venture capital funds.[149] In sum, these innovation policies stimulate the crossing of boundaries between academia and industry in several ways.

The creation of a hybrid research project

Conventional boundaries are increasingly challenged in late modern society resulting in the creation of hybrids: "Hybrids signal the breach of various socio-material categories, indicating inconsistencies that disorder routines and accepted mores" (Brown et al., 2006). As an academic-industrial collaboration the VIRGO consortium is a materialisation of this hybridity.[150] Science and innovation policy actively stimulate the crossing of boundaries between academia and industry in line with theories on innovation (Etzkowitz & Leydesdorff, 2000; Gibbons et al. 1994). Apart from academia and industry, government in fact plays an important role as third actor, implying VIRGO can be seen as a 'tri-brid', which resonates with the triple-helix theory. This theory explains how innovation takes place when the three domains of academia, government and industry become intertwined, especially when it concerns new techno-scientific fields like the contemporary biosciences.

---

[146] For the LINK programme, see http://www.bbsrc.ac.uk/business/knowledge/link.html, retrieved June 21, 2008.

[147] For Biopolis see Knapen, 2007 and http://www.biomed-singapore.com, retrieved June 21, 2008. Citation retrieved June 21, 2008 from http://www.a-star.edu.sg/biopolis/9-Biopolis.

[148] See, respectively, http://www.mediconvalley.com; http://www.cellulosevalley.com.au; http://www.foodvalley.nl; and http://www.biospace.com, retrieved June 21, 2008.

[149] For the analysis of the commercialisation of genetic knowledge and patenting, see, amongst many others, Rifkin (1999); Sterckx (1992); Thackray (1998) and Waldby & Mitschell (2006).

[150] For a quantitative analysis of science becoming a hybrid enterprise, see Owen-Smith (2003).

This section will analyse the building of VIRGO, starting with the academic origin of the project and efforts to turn it into big science. Next the Netherlands Genomics Initiative enters the stage, as it played an important role in shaping the VIRGO consortium. Finally, I will show how the VIRGO consortium turned into a collaboration between academia and industry.

An academic start

On the website of the Netherlands Genomics Initiative the VIRGO consortium is presented as a so-called 'Innovative Cluster', which means that the research is formulated in response to a question from industry and that industry takes the lead in the organisation of research (Folstar, 2002: 3). In the case of VIRGO, the leading company is ViroNovative BV. So when I decided to investigate the VIRGO consortium, I assumed that the main person behind the project would be someone from this company. However, it soon turned out that in order to get to know something about the consortium I needed to get in touch with coordinator Dr Arno Andeweg, an academic researcher who is based in the group of Osterhaus at Erasmus Medical Centre, which is part of Erasmus University in Rotterdam, a public facility. So while the coordination of VIRGO is presented as an industrial affair, it has a basis in academia.

As it turns out, the research involved started out as an idea of Andeweg. He has a background in biology and already began to be interested in infectious diseases during his studies. In his PhD research he focused on the Human Immunodeficiency Virus (HIV) under supervision of Osterhaus. After his graduation in 1995 he kept in touch with Osterhaus while working at several other research institutes where he became interested in genomics research. At the beginning of the new millennium he returned to the group of Osterhaus: "That is when I wanted to start this research" (interview Andeweg, 2005). He envisioned to integrate genomics research into the study of virus infections in order to learn more about the interaction between host and virus. VIRGO uses genomics techniques to investigate host-virus interactions to improve the rational design of vaccination and other intervention strategies for respiratory virus infections like influenza.[151]

The basic idea behind vaccines is that they prevent virus infections by artificially bringing the host into contact with the virus and learning the host how to react without becoming ill. However, vaccines can produce the good learning reaction, but also a bad and unwanted reaction and the crux is to know

---

[151] Retrieved October 29, 2007 from http//:www.virgo.nl

what makes the good reaction. However, Andeweg explains how up till now the reaction is often a surprise:

> We now basically do not have enough knowledge about the immune response of the host in case of a virus infection. So vaccine development is still largely depending on a 'trial and error' approach. If an experimental vaccine works we have a new vaccine, but if it does not work we have to try something else again. (interview Andeweg, 2005)

However, genomics research can contribute to the understanding of the host response to a virus:

> With the new genomics tools you can at every moment – this is like the time-axe within the black-box – and at each stage see which genes are turned on and which are turned off. With the new tools you now have the ability to look with a very high resolution into what exactly happens within the host and what happens if you change something in the virus or in the vaccine. (idem)

This gives better insight into the reaction of the host and can eventually rationalise the design of vaccines.

This new research approach had a slow start as Andeweg first had to work on existing projects, but he also got some time to work on his own ideas. When a European project was granted, he secured part of it to start his own research on a small scale. However, soon he realised he needed a large-scale approach:

> Genomics is big and technology development goes fast, so you actually cannot do this on a small-scale. This means that you need to realise, and this is my experience, that with only little money and little manpower you are always behind. And although it has your interest and it has potential, you will not be able to follow. (interview Andeweg, 2005)

Consequently, he tried to acquire more money. He first became part of a larger effort within the Erasmus Centre to become a 'Centre of Excellence' specialised in infection diseases within the newly established Netherlands Genomics Initiative. Unfortunately, they were not selected due to a lack of focus: "That was not very surprising because they involved more and more groups to fulfil the requirement of a very large multi-disciplinary centre. Everything may seem fit in the end, but if you do not take care, you lose your focus" (idem). When the NGI came with a new call for research proposals Andeweg decided to give it another go and write his own proposal. However, the research proposal needed to be aligned with the goals and requirements of the Netherlands Genomics Initiative.

During a political debate in the Netherlands in the context of the 'genomics revolution' and developments in biotechnology it was concluded that the Netherlands' genomics infrastructure was in need of a 'substantial reinforcement' (NGI, 2001). The lobbying of scientists and policymakers and the advice of a special committee[152] led to the establishment of the Netherlands Genomics Initiative as a special agency with the objective to stimulate and coordinate the genomics knowledge infrastructure in the Netherlands.[153] With the support of five Dutch ministries – involved with, respectively, education and research, health, economic affairs, agriculture and the environment[154] – the NGI was established in 2001 with a budget of € 188.8 million. The NGI became a special agency within the general Netherlands Organisation for Research Funding and was first directed by Professor Peter Folstar, holding a chair in knowledge management and innovation processes in the food industry at Wageningen University.

The first activity of the NGI was the creation of a strategy, which resulted in 'The Netherlands Genomics Strategy; strategic plan 2002-2006':

> The Netherlands Genomics Initiative heads the decision-making process for the selection and stimulation of both existing and new research activities. It primarily supports an integrated approach, from fundamental research up to and including ultimate application and attention to societal aspects. Significant emphasis is also placed on the education of young people and the positioning of genomics in social, national and international spheres. (NGI, 2001: 3)

The strategy of the five-year initiative revolved around the word 'focus': focus on excellency, focus on social awareness and accountability and focus on innovative potential. Dr Bernard de Geus, policy officer at the NGI from its inception in 2002 and responsible for project development, recalls how this strategy was created in a hurry:

> After the start of the initiative we were under great pressure. Everything had to be realised very soon, better today then tomorrow (…) The expectations were

---

[152] In 2001 the Wijffels Committee produced the report on 'knowledge infrastructure genomics', commissioned by the Ministry of Education, Culture and Science.

[153] The NGI was established in addition to already existing research funding organisations for life science research in the Netherlands; the Netherlands Council for Scientific Research has a branch for earth and life sciences (I-ALW) and biomedical research is funded by ZonMW.

[154] Education, Culture and Science; Economic affairs; Agriculture, Nature and Food Quality; Health, Welfare and Sport; Housing, Spatial Planning and the Environment.

high and we needed to start working immediately (...) When looking back at the process now, I think perhaps it would have been better if we had withdrawn ourselves behind closed doors for six months to reflect on how to organise things and how to create the best circumstances for research. (interview De Geus, 2005) [155]

The strategy was put into practice via an integrative approach consisting of twelve lines of action, most importantly the creation of 'Genomics Centres of Excellence' dedicated to fundamental research.

Overview of NGI activities in 2004 (NGI, 2005b)[156]

[155] In June 2007 De Geus accepted a position as director of the Top Institute Green Genetics, a new Dutch governmental initiative on agro-food biotechnology, News@genomics, June 5, 2007.
[156] Image courtesy of Netherlands Genomics Initiative.

The creation of 'Innovative Clusters' took place after the establishment of the Centres of Excellence. "The idea of the creation of Innovative Clusters emerged within the Initiative itself, because in the Centres of Excellence industrial participation was missing" (interview De Geus, 2005). Since the commercialisation of research is experienced as a difficult process in the Netherlands, the NGI developed a valorisation policy to realise innovation in the life sciences, and the Innovative Clusters are an important part of this policy.[157] By explicitly giving industry the lead in research, the Innovative Clusters are the materialisation of new insights in innovation theory that picture innovation as a cyclical process, instead of a linear development from academia to industry (Berkhout, 2002). The Innovative Clusters were not financed by the NGI, but from Bsik funds – a funding programme that supports the transition towards a knowledge economy with revenues from the old economy: the exploitation of natural gas. [158] Under supervision of the NGI, seven proposals devoted to the life sciences were prepared and submitted to the Bsik programme. Of this so-called 'NGI omnibus proposal' six proposals were successful – amongst others the VIRGO consortium – and they were granted a total of € 86 million euro of the Bsik funds.

While the NGI was originally established for a period of five years, efforts to prolong the initiative with another five years started towards the end of the period. In the first five years the NGI has built a network of genomics research in the Netherlands:

> We have been able to make some changes. Parties are starting to organise themselves, take their responsibilities and this can only be understood as the result of our actions (…) We have effectively stirred things up. (interview De Geus, 2005)

---

[157] Valorisation is the policy word for 'making money from knowledge'. Or as Horning, life sciences policy officer of the Ministry of Economic Affairs, explains: "Valorisation is the creation of value, or, when obtained in the context of (genomics) research: realisation of added value to obtained knowledge. This can be measured in economical terms by profit, high-quality employment and domestic product" (Horning, 2005). Next to the Innovative Clusters the valorisation programme of the NGI entails preparing and registering intellectual property, licensing intellectual property to existing organisations and business start-ups, supporting business start-ups and stimulating venture capital. In these efforts the NGI collaborates with organisations such as Senter-Novem, the technology foundation STW and Biopartner, who are dedicated to translational research, tech-transfer and the creation of new biotechnology companies.

[158] Bsik is an abbrevation of 'Besluit Subsidie Investeringen Kennisinfrastructuur' which translates into 'Decision Funding Investment Knowledge infrastructure'. Retrieved November 2, 2007, from http://www.senternovem.nl/bsik/algemeen/index.asp.

This view was confirmed by an international panel of experts during the mid-term review, which also brought the issue of continuity to the fore: "The panel was impressed by what has been achieved, but also expressed the view that more work will be required to achieve the desired and intended objectives" (NGI, 2006: 32). Based on an overview of results over the period 2002-2007 and a new strategy for the period 2008-2012 supported by industry (NGI, 2005a; NGI, 2005c; NGI, 2005d), the Dutch government decided in September 2007 to award the NGI € 271 million for a second phase (NGI, 2007a). In the coming years the NGI will concentrate especially on maximizing the economic and societal value of the research as presented in the new NGI Business Plan which is therefore called 'Munt uit Genomics', which translates as 'Capitalising on Genomics' (NGI, 2007b).

Building an 'Innovative Cluster'

The Netherlands Genomics Initiative has played a crucial role in the building of VIRGO. After the first Rotterdam proposal for a NGI Center of Excellence in which Andeweg and his research would have a place was declined, Andeweg decided to give it a go on his own when the NGI came with a new call for proposals for the Innovative Clusters: "it gave me the space to do what I was convinced of together with Osterhaus as a motivating factor" (interview Andeweg, 2005). It took many of his Sundays and much more time to work on the research plan:

> Setting up a research project is a big investment for a researcher and a research group, so you have to be able to find the time to actually participate in the competitions that enable you to scale-up your research. Personally, I created that time by more or less putting my experimental work on halt when my lab-analyst went on maternity leave. So I did not have to be in the lab that often and I could dedicate myself entirely to the research proposal. (idem)

Based on previous experience with writing research proposals, Andeweg made sure he kept his plans focussed. For instance, he decided to concentrate only on respiratory viruses. These are different kind of viruses and therefore provide the necessary variety, while they are at the same time related.[159] Andeweg did not organise meetings with collaborators but mainly communicated via email or

---

[159] The viruses that are targeted are the influenza virus, the Respiratory Syncytial Virus (RSV) and the human MetaPneumoVirus (hMPV).

held bi-lateral meetings and only after designing the research proposal other research groups where selected to be part of the research project.

In addition, Andeweg was well aware of the importance of presenting collaboration and made the proposal readable and attractive for the evaluators. For instance, he made sure to include diagrams in the research proposal.

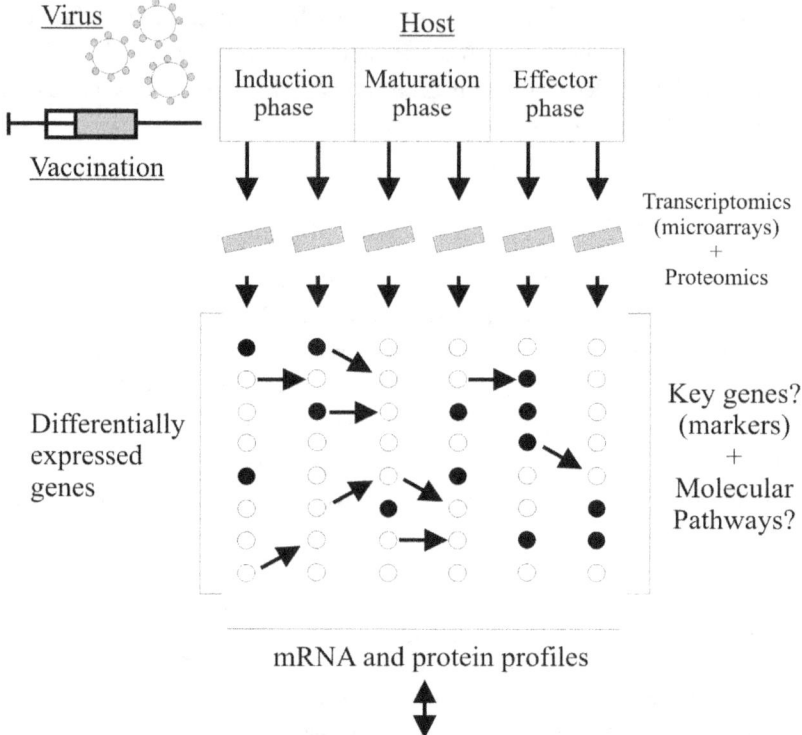

The schematic presentation of the application of
genomics tools in research on the virus-host interaction[160]

I learned from the EU project in which I participated that they like pictures. This is simply what works and it is not artificial as I also like to think along these lines. When people have to read lots of this kind of paperwork [he flips

---

[160] Image courtesy of Arno Andeweg, VIRGO.

through the elaborated research proposal], then it is important that you show with nice diagrams how the structure is built. (interview Andeweg, 2005)

Finally, Andeweg and Osterhaus baptised the collaboration VIRGO: a contraction of virology and genomics which also indicates that the research has not been done earlier.

In February 2003 the VIRGO research proposal was submitted and in the following months reviewed by several organisations:

> From February till the summer numerous organisations have taken a look at it, from the Netherlands Bureau for Economic Policy Analysis to the National Institute for Public Health and the Environment and SenterNovem for economic-societal factors and of course the purely scientific review as well. (interview Andeweg, 2005)

The VIRGO team had to present the proposal four times in a row to different organisations, but it was worth the trouble. In November 2003 their efforts were rewarded with a grant of € 10.8 million (NGI, 2003: 11).

But what about the participation of industry? When asking Andeweg about this, he acknowledges that the requirements for VIRGO were quite different from the requirements of the academic Centres of Excellence of the NGI. For instance, in an Innovative Cluster, industry needs to be the leading party. Consequently, the VIRGO consortium put forward ViroNovative, a company that is dedicated to the human metapneumo virus (hMPV) which was discovered by the group of Osterhaus and first published in 2001. This resulted in the establishment of a new company that is located in the same building as the group of Osterhaus, only two floors higher:

> This company is a spin-off of our group. So that was easy, as Ab [Osterhaus] is the scientific director of the company. We could perfectly combine the two so as to find a good solution. And we all work both above and here below anyway. This good 'internal' public-private collaboration, then, provided the foundation for the expansion of public-private collaboration in the innovative cluster. (interview Andeweg, 2005)

According to Andeweg, the NGI knows that ViroNovative is not really the leading party behind the consortium, but this is not a problem as VIRGO is one of the clusters with most commercial ties.

When it comes to the formulation of the question by industry that clearly did not happen in case of VIRGO, Andeweg states:

It depends on how you define what a question is. Pharma in general and Solvay [one of the participating companies] in particular, have a great interest in better vaccines, so there is a demand. It is only that they did not ask us explicitly to do this using genomics tools. (interview Andeweg, 2005)

The industrial partners involved have different aims. ViroNovative wants to exploit its IP position further by developing knowledge for marketable products for hMPV intervention strategies, while Solvay Pharmaceuticals is interested in the development of knowledge for a new generation of human influenza vaccines. Next to ViroNovative and Solvay, Intervet takes part in the project, aiming to develop knowledge for animal influenza and APV vaccines. The different orientations of the companies involved, enables them to build on the same knowledge without competing with each other.

Academic and industrial partners agree on being complementary. Coordinator Andeweg acknowledges the importance of academia and industry working together, as it is important that academic research on vaccines has an outlet to companies. In addition, the group of Osterhaus has a long-term tradition of cooperation with industry to supplement governmental research funding which is 'not always abundant'. The other way around, companies depend on public research as they are often not able to perform fundamental research themselves. The industrial partners hope to acquire new fundamental knowledge through the project that can be further developed into marketable products. Jeroen Medema from Solvay Pharmaceuticals explains how VIRGO contributes to vaccine development.

The VIRGO consortium is of great importance for vaccine producers. First of all, we do not have the capacity to carry out that kind of fundamental research ourselves. Our strategy is to monitor what is going on in universities and elsewhere and then jump on the bandwagon, preferably before it accumulated speed. That means we try and pick up new concepts at an early stage and develop them further through clinical trials, registration and market introduction. We expect VIRGO to feed our pipeline of new vaccines and medicines. (Medema cited in NGI, 2005e)

Medema especially expects to use the research in VIRGO to improve vaccines and reduce the time that is needed to develop a vaccine.

While the commercial parties are now seriously involved in the research and contributed € 3.3 million to the project (Boekholt et al., 2007), the formation of VIRGO shows how the research project is actually an academic collaboration that turned into a public-private collaboration led by a company due to the requirements of funding policy. This resulted in the VIRGO project proposal in which actors brought their different goals into line by combining the fundamental research with industrial goals for application. Nevertheless, the academic partners still have primarily academic goals: the development of new knowledge about host-virus interaction on which the participating academic groups can build in future research.

The project as a boundary technology

The former section showed how in interaction with government, an academic-industrial collaboration was created. Thereby the project combined the different agenda's of academia, government and industry; while scientists seek money to develop a new scientific approach, government wants to stimulate innovation to improve the economy and industry needs new knowledge to develop new vaccinations that will bring profits. The specific agendas are – together with specific structures, procedures and timeframes – fundamental to the stories and actions of each specific domain (Law, 1994). These different 'modes of ordering' need to be interwoven in the 'tri-brid' research project VIRGO. This section will show how the creation of VIRGO in which three societal domains become entangled, requires not only the creation of a common objective, but also the connection of the different modes of ordering of academia, government and industry.

To investigate how the three different domains grew intertwined, I will explore how their different modes of ordering interacted to create the VIRGO project. The project format has become an important means to structure science. As projects are employed in both industry, government and science, it is a way of working that the three domains have in common. Within the project format, tensions between different modes of ordering are resolved, or at least a workable solution is found. The project links various organisational structures, but also orchestrates different procedures. Think of agreements on research materials and technologies used within the project, but also the division of the research budget and ways to allocate IPR are agreed on in the project proposal and later within the project management. This means the project can be seen as a 'boundary technology' that enables the connection of government, academia and industry.[161] The project format serves as a tool which facilitates this connection work by enabling the crossing of boundaries between the domains. If

---

[161] For the boundary concept see Gieryn (1983, 1995, 1999); Star & Griesemer (1998); Guston (2001).

the project is successfully set up it functions as a 'boundary organisation', an organisation located at the boundaries of different societal domains.

The structure of VIRGO

Next to the creation of a common goal, the structure of VIRGO makes visible how academia, government and industry have become aligned. First of all, the research project is connected to the Dutch government via its funding sources: the Netherlands Genomic Initiative and the Bsik funding programme. This last program is a collaboration between six Dutch government departments, including the Ministry of Education, Culture and Science which is responsible for the Innovative Clusters of the NGI. However, the monitoring of the Bsik programmes is delegated to SenterNovem, an agency of the Dutch Ministry of Economics Affairs that stimulates innovation. As a result the VIRGO consortium is officially tied to two funding programmes and three government organisations. Moreover, the requirements of the funding organisations have connected the academic partners to the industrial partners in the collaboration.

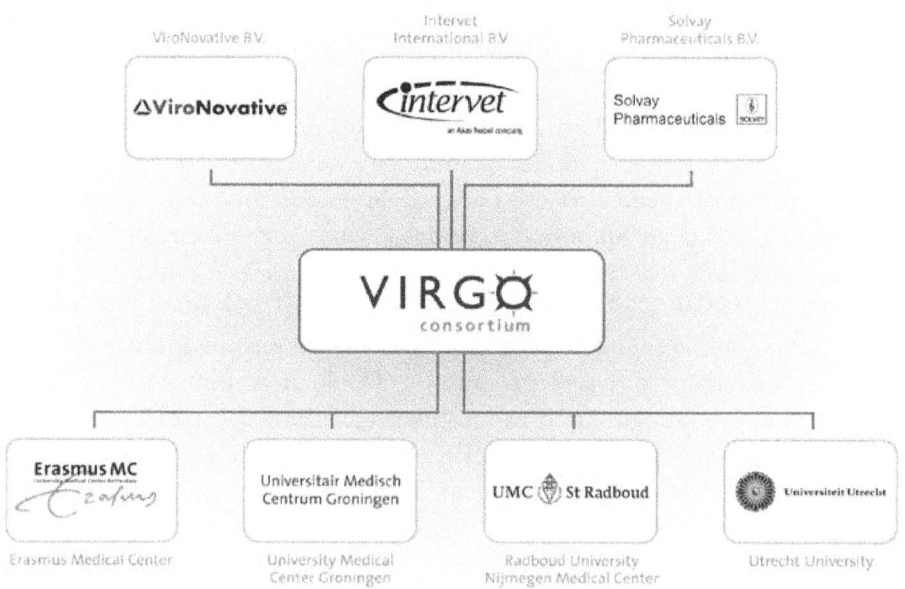

Partners in the VIRGO consortium[162]

---

162 Image courtesy of Arno Andeweg, VIRGO.

The VIRGO consortium itself consists of research groups from four universities and three pharmaceutical companies. These partners were already connected to the group of Osterhaus in some way. Next to the spin-off company ViroNovative, that has become part of the American company MedImmune, Solvay Pharmaceuticals is already a long-term partner of the virology department in Rotterdam regarding research into human influenza.[163] In addition, Intervet, specialising in animal pharmaceuticals, was already involved in genomics research, notably related to poultry.[164] The academic partners involved in VIRGO are colleagues of Andeweg dispersed over four Dutch universities. They specialise in virology, theoretical biology, medical microbiology, veterinarian medicine, pulmonary medicine, immunology, paediatrics, neurology, and bioinformatics. In this way, the VIRGO project solidifies already existing connections.

Within this organisational structure, the research is divided into ten so-called 'Work Packages': "Of each WP we elaborated on the participants, the goals, the approach, the detailed approach, the milestones and the responsibilities" (interview Andeweg, 2005). Each work packages is performing a specific part of the overall research plan: that study respectively host-virus interactions in target cells (1), mouse models (2), non-human primates (3), chickens (4) and humans (5) (Johnston et al., 2007). Other packages are dedicated to vaccine research (6), transcriptomics (7), proteomics (8), and data storage, analysis, mining (9) and the modelling of immune gene-interacting networks (10).

While the Work Packages are performing the research independently and each has its own research leader, they are related to each other through the overall aims of the research and the fact that they also build upon each other's work. For instance, the cell and animal studies are performed to find basic mechanisms of host-virus interaction that may also take place in humans. Therefore, Work Package 3, which uses non-human primates for experimentation, explicitly builds on the work within WP 1, 2 and 4 in which respectively cells, mice and chicken are used as a model. WP 6, which performs research on vaccination, again builds on WP 1 to 4. In addition, WP 7 is dedicated to technology development and standardisation for all projects and partners in the consortium, which is crucial for making data compatible. Finally, WP 9 and 10 integrate the data from all the other research efforts and build models.

---

163 More information about the company can be found at its websites. Retrieved, November 2, 2007, from http://www.solvaypharmaceuticals.nl; http://www.solvaypharmaceuticals.com
164 More information about the company can be found at its websites. Retrieved, November 2, 2007, from http://www.intervet.nl; http://www.intervet.com/

Next to the individual management of the Work Packages, an overall management structure of the VIRGO consortium is in place, consisting of a general assembly, a steering board and a kind of governing board that reports to Senter and the NGI. Each WP has a representative in the general assembly who at the same time represents one of the participating institutions. The steering board is responsible for daily business which is executed by Andeweg in dialogue with Osterhaus. In addition, the project is supported by a half-time secretary and some people with business experience are involved. The project is making use of the experience of an officer of the Technology Transfer Office of Erasmus MC to allocate IPR. Finally, legal and financial matters are outsourced to experts, respectively to the legal service of Erasmus MC and an outside accountancy firm.

The structure of VIRGO thus shows how the different domains are tied together in an organisational structure which has various layers. First of all, there is a core –the VIRGO consortium – consisting of academic and industrial research groups. Around this centre, government agencies are in place, as well as financial and legal support. Secondly, the project itself has various organisational levels. While at first sight the fourteen different research groups all seem to collaborate, it turns out that the research is actually divided into ten Work Packages in which specific groups work together. These groups are connected through the research results and via the management of the project. It is only on the management level that actors from all the different organisations come together. This structure binds the different domains together in a new organizational entity: the VIRGO project. As a result the project format forms the backbone of VIRGO as it enables the different societal orderings to connect and become aligned.

Dealing with bureaucracy

Next to structuring the research, the project deals with bureaucracy and accountability. When talking to scientists who are in charge of a research project, you can be sure that at some point they get started on 'bureaucracy', a term that scientists use to cover the things they certainly do not like: policy procedures and the pile of paperwork that comes with it. Originally bureaucracy does not have a negative connotation and it can be argued that bureaucracy is the protector of freedom in scientific collaboration as it defines the participants' rights concerning data (Shrum, et al., 2007). Nevertheless, for scientists bureaucracy is a major cause of frustration and the direct enemy of valuable research time. In this respect Andeweg is exemplary when claiming to prefer that government just hands out a bag of money without restrictions:

Sometimes I wonder what happens if you just give researchers money and a direction of research and let them do their job. Of course things will go wrong, but the question is whether more things will go wrong? At least it would mean that more money is spent on research as now the costs for the whole apparatus are quite substantial, not only at the policy side, but also at the academic side. (interview Andeweg, 2005)

In this light, the need for an NGI was even questioned by scientists. For instance, the prominent Professor Piet Borst, former director of the Netherlands Cancer Institute, claimed:

Already before the establishment of the NGI, excellent research in genomics was performed in the Netherlands. Researchers only needed money to be able to perform world-class research, but they did not need orchestration. They know themselves which research directions are promising. They needed help, not interference or extra bureaucracy. (Borst, 2004: 46)

From a scientific perspective the new initiative with its elaborate strategies, fancy brochures and network meetings seems a waste of money, because it is the research that counts and that is where the money has to go.

A different perspective, however, also shows a different world. From a policy perspective De Geus from the NGI states that the regulations and paperwork that come with the funding of science are simply essential:

The Netherlands Genomics Initiative is viewed as quite bureaucratic indeed, but this is all but true. Yes, we do have some rules people have to adhere to. But when we ask for reports of progress we want to know only about the general progress, not the details. Rather than lengthy reports of progress, we want concise ones. Accountability is the real problem, however. People just don't like to be accountable, especially scientists. But accountability is not a strange thing to ask for. When I award a research grant of some 16 million, I would think I'm entitled to know what happens with the money. (...) We are talking about public money here that should be accounted for. (interview De Geus, 2005)

On a European level, Dr Jacques Remacle from the Directorate General F4 specializing in functional genomics, adds that the need for accountability increases together with the scale of research:

When dividing money over small research projects, it does not matter if one project does not deliver as the others will. However, when investing huge amounts of money in large research networks I need to know how the money is spent. (interview Remacle, 2005)

Moreover, it is argued that research policy takes place on a playing field in which science is not the only player, and that research has to compete with other national priorities and should produce tangible economic or social benefits.

These dissimilar scientific and policy perspectives on bureaucracy are to some extent reconciled in the VIRGO project. The project structure enables the scientists to develop their own internal management practices, while also staying accountable to the funding organisation. On the one hand, the internal coordination of the project can be organised by participating scientists, minimising bureaucracy. Within the VIRGO project they explicitly tried to keep the organisational structure simple:

> In other consortiums they put an extra management layer in between, but they often take the bureaucracy of The Hague [the seat of government and many funding organisations] into the research projects (…) In contrast, we are very decisive compared to the sister projects who often have about three people being responsible for daily business. (interview Andeweg, 2005)

Within VIRGO they chose for one central coordinator. Andeweg is like a spider-in-the-web, being involved in research as well as management and communication towards outside organisations. Although participants sometimes get annoyed when Andeweg sends them too many different emails, they certainly appreciate the lack of bureaucracy.[165]

On the other hand, the project format enables accountability and evaluation of science by making science open to external control. By constructing VIRGO as a research project, it becomes a separate organisational entity with pre-set goals that can be evaluated, not only at the beginning and the end of projects, but also at regular intervals in between. In case of VIRGO three different evaluation procedures are in place. VIRGO started with a so-called 'zero measurement' in which the situation at the beginning of the project is pictured. This measurement is followed by monitoring halfway and at the end. Secondly, reports of progress for the Netherlands Genomics Initiative have to be made every six months. In addition, VIRGO is part of evaluations of the NGI as a whole. Moreover, under the influence of government and industry standards of evaluation become more diverse in comparison to the peer-review that is the common form of evaluation in the scientific domain. Reports of

---

[165] In the beginning of 2009 Andeweg informed me that the VIRGO consortium also had to expand the management of the project by hiring special project managers (personal communication with Arno Andeweg, January 12, 2009).

progress not only focus on scientific results but also on the management process and societal evaluation criteria. In addition, evaluations pay attention to the commercialisation of research results, such as the number of partners, patents and start-up companies.

Making time and space

Next to dealing with bureaucracy and accountability, the project enables the creation of a project time and space. Firstly, the project intermediates between the different time regimes in science, government and industry. While research results can take quite some time and certainly do not come at a pre-set time, administrative time has an annual rhythm and is relatively short-term. In addition, industrial time is configured as 'time to market' (TTM), which refers to the time it takes to transform knowledge into a product that can be sold and which is ideally as short as possible. The research project goes beyond these different orderings by making its own time. This coordination of different temporal regimes can be clearly seen at the start of VIRGO. While the development of the new line of research by Andeweg already started at the end of the last century, the project proposal for VIRGO was handed in at the beginning of 2003. Although they soon heard that chances of funding were high, and funding was officially confirmed in October 2003, the year 2004 was well under way before they could officially start the research:

> In the summer [of 2003] we already knew that we were in second position concerning the science review of about seventy projects that would eventually get funded, so we knew we had a very big chance. But it took almost a year before we actually got the letter that we could start. That was around March or April 2004. And to make matters worse, we had to start retro-actively in January. (interview Andeweg, 2005)

So while the research project could only start after they received the letter, the official starting point of the project became the beginning of a new year: January 2004.

Although the project format enables harmonisation of different time-frames it proves difficult to align them smoothly. This can for instance be seen in the process of designating a common starting point, but also in the evaluation procedures. As January 2004 was considered to be the project's starting point while the research had not actually started yet, the first six-month progress reports posed a problem because no research took place yet. A similar problem became apparent during the external evaluation of the Netherlands Genomics Institute in 2006. Four years after the start of the NGI, the initiative was subject to thorough evaluation, including the eleven research centres that were established under its supervision. As a result, the VIRGO consortium was evaluated by an international review committee chaired by Sebastian Johnston, Professor of Respiratory Medicine at Imperial College in London. Although overall the consortium was evaluated as 'very good to excellent' the report states:

> This evaluation is carried out prematurely as the Consortium only received confirmation of its funding in October 2004 and many of the Work Packages only started their work between 1 year and 1½ years ago. The Committee were really only able to review planned activities and preliminary data. Productivity for all Work Packages was impossible to assess. The scoring of the various Work Packages has potential to be considerably higher than that awarded in this assessment as there was little in the way of outputs available for review. The 'Work in Progress' was generally considered to be of excellent quality. (Johnston et al., 2007: 11)

So Professor Johnston and his colleagues from Canada, Northern Ireland and Spain performed a review of the VIRGO consortium in which they concluded that the time was not right for an evaluation yet as the research only just started.

In addition to time, different spatial orderings need to be aligned. While science takes place in international community and industry is often multi-national and operates on global markets, government has a national orientation. Because within the VIRGO project these different spatial orderings are re-aligned into a national space, the creation of VIRGO shows how sometimes a specific ordering can be more dominant in the new project configuration then others. While flu research is an international activity and the group of Oster-haus works together with various international partners, VIRGO consists only of Dutch research institutes and companies located in the Netherlands. The funding sources are the reason that VIRGO is a national collaboration. The NGI has been established to build a genomics research infrastructure within the Netherlands and does not support scientists from other countries and Bsik money also has a national label. Only when the NGI broadened its scope to include internationalisation as an objective, the international context of the VIRGO project became more prominent. Consequently, the spatial ordering of government proved to be authoritarian within VIRGO.

Hybrid scientists

VIRGO does not only have a hybrid structure, its creation also asks for hybrid scientists: scientists who are able to combine doing science with research management or the commercialization of research. When reflecting on the construction of VIRGO, Andeweg notes that building collaboration is a different way of doing science, as coordination and management become more important: "The motivation to start such a project is the fascination for the content of the research, but in practice you gradually transform into a manager" (interview Andeweg, 2005). Managing research can have important consequences for the career of a scientist. Building a new research project simply affects the academic performance of a scientist, as no time is left for performing research and writing publications:

> I just got to the point in this research project where I can start thinking of publishing again, but I didn't have that for years. Of course it does matter if you manage to acquire such a project, because it allows you to stay on a bit longer. But I made an enormous time investment that might as well have left me with nothing if the project would not have been funded. It is victory or death. (idem)

Obviously, setting up and managing projects requires a huge investment of time from scientists that cannot be spent on research or writing publications.

In addition to the lure of management, the crossing of the boundary between academia and industry has implications on an individual level as well. For instance, within the Industrial Cluster format of VIRGO it would have been easier if Andeweg as the person in charge had started to work for the leading company ViroNovative. However, he explicitly refused to make the transfer to industry as he prefers being an academic:

> I simply do not want that. You will get the most bizarre situations, for it is all about money there, and you are also dependent on the decisions of the American parent company (...) My heart is with research and I want to do that within academia because I want to be independent. (interview Andeweg, 2005)

At the same time Andeweg notes how being an academic increasingly can be compared to being an entrepreneur. Scientists can only perform research when they write research proposals. If in this case Andeweg made a clear choice to stay within academia, his role as an academic transformed anyway. He is forced to balance managerial interests and scientific standards: a double role.[166]

Although Andeweg stays put, individual border crossings have become quite common within the scientific world these days. Some scientists even seem to be particularly good at it: they manage to strike a balance between science, business and policy interests and in addition sometimes even master public communication. A famous example is Craig Venter who, after the beginning of his scientific career within the National Institutes of Health, reinvented himself as an entrepreneur competing with the public project to sequence the human genome (Shreeve, 2004b; Venter 2007). If this made him world famous, it also turned him into a hybrid scientist.

BusinessWeek

DECEMBER 13, 2004
THE GREAT INNOVATORS

Craig Venter: DNA's Mapmaker
The scientist's maverick approach opened the floodgates of genomic information

By John Carey

Craig Venter as pictured in Businessweek[167]

---

[166] See Packer & Webster (1996) for an analysis of the way in which scientists move between academic and patent cultures to manage their rewards.
[167] Image courtesy of Gregory Heisler.

174

The official leader of VIRGO Ab Osterhaus performs a similar hybrid role within the Dutch national context and the international community of influenza experts. He combines his role as a successful academic, with his roles as director of the Dutch National Influenza Centre, government advisor and entrepreneur. In addition, he regularly appears as an expert in all kinds of media to talk about the risks of influenza.

Although Osterhaus' embodiment of different identities has been an asset for the promotion of VIRGO in the different domains, the combination of different roles can also give rise to criticism. In the case of Osterhaus it was the epidemiologist Dr Luc Bonneux from the Belgium Health Care Centrum who basically challenged the different roles Osterhaus combines (De Rijck, 2005).[168] Bonneux took a stance against virologists like Osterhaus who are predicting the coming of a pandemic and advising governments to buy anti-viral medicines. According to Bonneux there is not enough knowledge of influenza pandemics to predict the coming of a new one. In addition, if the risk of a pandemic would be real, anti-viral medicines are a questionable solution that only certainly benefits one party: the pharmaceutical industry which produces the anti-virals. According to Bonneux the money governments spend on anti-viral medicines can better be invested in the improvement of the health system at large. In short, Bonneux suggests that some virologists do not only want to improve health safety, but also want to secure money for their field of research and benefit companies they are working with. Thereby he basically questions whether it is possible to combine the different identities in a single person, suggesting that a scientist should keep to his scientific role.

Conclusion

When taking a closer look at collaboration between academia, industry and government, it becomes clear it involves a complex phenomenon. The three different domains bring their own ordering of the world, and together they have to compose a new configuration. My analysis of VIRGO demonstrates how the project format plays an important role in bringing academia, industry and government together. The project accommodates the crossing of boundar-

---

[168] This critique was first formulated in a newspaper article in the Belgium national newspaper *De Standaard* and was followed by an interview on Dutch radio in 'De Ochtenden' on October 19, 2005 (Retrieved December 1, 2005 from http://www.ochtenden.nl/afleveringen/23986305/) and a debate between Bonneux and Osterhaus on NOVA, a Dutch current affairs television programme, on October 22, 2005. (Retrieved December 1, 2005 from http://www.novatv.nl/-index.cfm?LN=nl&FUSEACTION=videoaudio.details&REPORTAGE_ID=3808).

ies through the creation of a shared goal, a new structure, common procedures and particular temporal and spatial orderings. However, the intermingling of the three different domains is not straightforward. Next to the crossing of boundaries between domains and the merging of different modes of ordering, the building of VIRGO also shows how boundaries become realigned or are kept steady. This can be seen for instance in the objectives of the consortium. Although the different parties involved have formulated a common goal, they maintain their own separate goals underneath the shared goal. In addition, the structure of collaboration maintains boundaries between different domains; while the overall structure of VIRGO presents collaboration, the actual research takes place in separate Work Packages that only require specific researchers to work together. In addition, scientists have made procedures inside the project as simple as possible, trying to keep governmental bureaucracy outside of daily project activities. Finally, borders are protected on an individual level. While individual scientists are able to cross boundaries between the different domains, some decide to stay put in the scientific domain. As a result, collaboration has a different face on the level of management, the actual research practice and the individual level of scientists.

The innovation epidemic

This chapter explicitly looked into the role of government and industry in building scientific collaboration. As in our current 'knowledge society' or 'knowledge economy' innovation is a measure of good governance, governments are trying to influence the inventiveness of their country trough the creation of networks – based on broad theories on innovation that observe the increasing importance of connections between academia and industry to create novel products and services (Barry, 2001; Etzokowitz & Leydesdorff, 2000; Gibbons et al., 2004). Within the realm of the life sciences, innovation policies clearly left their traces. The creation of VIRGO as an academic-industrial collaboration is part of life sciences research and innovation policies that have spread around the world. In other words, innovation takes place through imitation.

However, when looking more precisely at practices in which innovation takes place it becomes clear that invention also importantly resides in arrangements:

> inventiveness should not be equated with the development of novel artifacts, or indeed with novelty and innovation in general. Rather, inventiveness can be

viewed as an index of the degree in which an object or practice is associated with *opening up possibilities*. In this view, scientific and technical objects and practices are inventive precisely in so far as they are aligned with inventive ways of thinking and doing and configuring and reconfiguring relations with other actors. (Barry, 2001: 211)

In the case of VIRGO it became visible how scientists are inventive in opening up possibilities to get research funding and adapt their research and its organization to funding requirements.[169] But does this result in a really inventive arrangement of relations? Although, governmental policies to a certain extent allow the configuration of collaboration in biology to particular circumstances, this chapter also showed how scientific collaboration is subject to a worldwide innovation regime which leaves little room for inventive ways to configure and reconfigure relations in collaboration.

Finally, it can be observed that within VIRGO the scientific way of ordering becomes repeatedly secondary to other modes of ordering. This can be seen for instance in the spatial orientation of the project that follows the governmental mode of ordering and the evaluation of research that complies with the annual political rhythm while ignoring the pace of scientific developments. In addition, the objectives of the project shift from academic publications to results that fit into the industrial mode of ordering. On evaluation forms, the number of expected patents have to be filled in:

We sometimes have a good laugh about these forms. If you can predict what will be the result of your research, you do not have to perform the research anymore (...) And they want it [the number of patents] specified per year. It is like having to predict in which city you will live 10 years from now and also knowing in which street and at which number (...) If I already knew, I would not be working in academia but I would work as an adviser to a company. (interview Andeweg, 2005)

Tellingly, in such cases the intermingling of the three societal domains – science, industry, government – has more impact on scientific ordering than on the other orderings, a phenomenon called 'asymmetrical convergence' (Kleinman & Vallas, 2006).[170] In this respect my analysis of the creation of VIRGO underscores how scientific ordering is profoundly affected in the building of academic-industrial collaboration.

---

[169] For tailoring biology research to make it appear more applied see also Calvert (2006).

[170] This term is coined by Kleinman and Vallas to refer to the process of research cultures in science and industry becoming more similar these days, while industry affects scientific culture more than the other way around, which renders the process asymmetrical.

PART 3

Supersizing science

CHAPTER 6

Unpacking collaboration in biology
Supersizing science

How to collaborate in biology? While studies on scientific collaboration tend to be devoted to disciplines traditionally seen as collaborative, like physics, astronomy or space research (where the very objects of science are scale dependent and so encouraging of big science), I have turned the spotlight on scientific collaboration in biology. Although collaboration in biology is often pictured as a recent phenomenon, with the Human Genome Project as a turning point towards increasing – or, as some will say, necessary – collaboration, my argument has been geared towards demonstrating that biology does have quite a long history of collaboration. By looking back in time, I argued that the first forms of scientific collaboration can be found in the early natural history tradition. The gathering and classifying of the various divergent forms of life on earth has always presupposed a collective scientific undertaking rather than an individual one. The increasing prominence of laboratory biology, which in some ways added a more individualist dimension, has shaped the image of biology as a non-collaborative science, or at least as an undertaking that is usually limited to a local, small lab-based group. Currently this image is changing again, however, as laboratory biology increasingly takes place within large-scale collaborations, which in this thesis I have characterised as a networked form of big science. My concluding chapter will refine this characterisation by further unpacking large-scale collaboration in biology. In so doing I will first attend to the organisation of collaboration in biology and subsequently to the work involved in actually building collaboration: the supersizing of science. Finally, I will reflect on the meaning of the study of collaboration in biology for the study of scientific collaboration in general.

The organisation of collaboration in biology

A defining feature of scientific collaborations is their organisation. After explaining how collaboration in biology has a networked form, the analysis of three large-scale collaborations gave insight into the different ways in which these research networks are organised. This section will examine the organisation of collaboration in biology in more detail. How do collaborations vary in structure and size? I show how in the different case studies the form and character of research are related, resulting in different styles of collaboration in biology.

The structure of research collaboration

Overall, collaborations in biology have a similar networked structure. The Human Genome Project and my three case studies all have a decentralised structure held together by a central governing entity. The projects basically comprise different entities with their own leadership structure, topped by some sort of central management construction, which is often divided into scientific and organisational management. However, within this overall structure, differences can be located which are related to the different orientations of the collaborations. In general, scientific collaborations vary from formal to less formal and from more hierarchical to democratic. Looking at the division of labour, leadership, degree of formalisation and decision-making hierarchy, Shrum et al. (2007) distinguish three types of scientific collaboration: bureaucratic, semi-bureaucratic and participatory. Accordingly, it is possible to identify marked differences between the networked structure of the case studies: VIRGO can be characterized as bureaucratic, the Silicon Cell as semi-bureaucratic and the Census of Marine Life as participatory.

VIRGO is an example of a bureaucratic collaboration because it is most formally organised, with a clear division of labour and a management structure which includes a scientific leader and a programme coordinator. In contrast, the Census of Marine Life is much more participatory. Although there is a clear central management structure, including a scientific steering committee and a secretariat led by the projects senior scientist and a program manager who sets out the basic parameters, the different research programmes and entities have a lot of autonomy. While the central management provides the overall structure in which collaboration takes place, decision-making has a democratic character and coordination is also a collective activity. Finally, the Silicon Cell can be characterised as semi-bureaucratic. While at this point, research on the Silicon Cell and the later systems biology initiatives does not take place within any of the structures envisioned, the scientists involved were aiming for autonomy of participating groups with a central coordinating body in which agendas and standards are negotiated.

The character of research seems to influence the structure of collaboration. Projects are more bureaucratic when the interdependency between the collaborators is high and more participatory when researchers are less dependent on each other's work (see Figure 1). For instance, in the VIRGO project there is a high interdependency between collaborators because the different research units build upon each other's results. The data from some Work Packages are essential to the experiments in the other WP's, and results are finally integrated in the WP that builds a model. This interdependency is reflected in the bureaucratic structure in which relationships are clearly defined and responsibilities divided. Initiatives in systems biology have a similar kind of interdependency, even if it is less immediate. For instance, when wanting to build a Silicon Cell, research can be performed independently, but similar standards are needed in order to fit results together in the actual model. By contrast, when looking at the Census of Marine Life there is no real interdependency as the research is performed separately and only afterward the various results are uploaded in the uniform database. As a result, the Census offers a potentially very loosely organized frame.

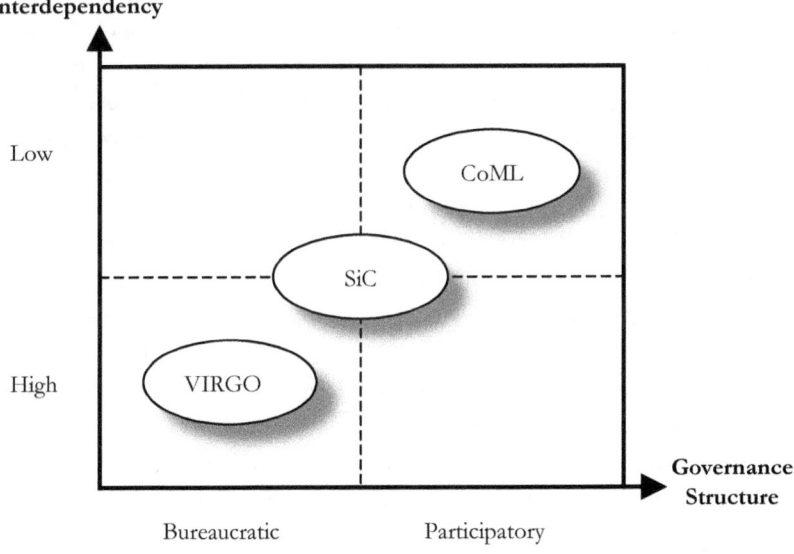

Figure 1: the relationship between the governance structure and
the interdependency of collaborators

In sum, projects that are merely geared towards collecting and ordering information are less interdependent and can therefore be more loosely organised than projects dedicated to the creation of models or applications that depend on *shared* material objects. Although a project such as the Census might as well have a more bureaucratic structure, it is difficult to imagine that a project such as VIRGO – in which research is highly integrated – takes place in the comparatively open structure of the Census project. In fact, this project's current problems with integrating research results seem closely linked up precisely with its loose organization.

The size of research collaboration

What is the size of big science? According to Shrum et al. (2007) the size of scientific collaboration in general can be measured in different ways, varying from the character of instrumentation; the scope of research; the number of people, organisations or teams participating; the funds required, available or utilised; the time required to get the project off the ground; and its duration. Because big biology has a networked structure, the geography of collaboration is crucial in determining size. What terrain does a network cover and how many countries are participating? The collaborations I have analysed vary in size: while VIRGO is a national project, the Silicon Cell Initiative wants to become European and the Census of Marine Life is an international project that aspires to become truly global. Although several factors interact in determining the size of actual research projects, my argument has elucidated how in every single case a different factor is critical in establishing size. This allows me to make a distinction between three different models of 'bigness': aspirational bigness, material bigness and policy-imposed bigness.

First of all, bigness can be aspirational, or actively pursued from the start. This can be seen at the beginning of projects, in particular in the case of the Silicon Cell initiative. Here the research is small-scale, while the initiators want it to be big, which gives rise to a sense of promise of bigness. However, different arguments are used to legitimise this aspiration, including material bigness. The material dimension of bigness can be differentiated into bigness related to technologies, resources, data or standardisation. The international Census of Marine Life involves a resources related material bigness. As life in the oceans is dispersed globally, the project seeks to involve as many ocean-bordering countries as possible. Next to the objectives of research and the research material, also research policy and funding structures can determine the size of collaboration. This policy-imposed form of bigness is visible in particular in VIRGO, which is a national research project while virus research is an international field of research. The project could have been set up as an international effort of course, but the funding format of the Netherlands Genomics Initiative simply did not allow for that and as such it had a decisive influence on the project's size.

This restrictive policy effect on the size of collaboration also nicely illustrates how national funding sources may run counter to or even frustrate the accommodation of international collaboration. This same dynamic is visible in my two other case studies as well. The Silicon Cell Initiative and the Census of Marine Life are not possible with only national funding sources, as nations typically invest in scientists associated with their country, rather than in scientists in faraway countries. Although new international spaces for research collaboration are built – for instance in the European Union – and scientific collaboration is often stimulated by science policy and funding organisations, these actors can restrict international collaboration at the same time. The Census solves this problem by a patchwork structure: nations paying the actual research of their scientists, while the Sloan foundation is providing the money for the international coordination work. In turn, the scientists of the Silicon Cell try to overcome the national funding borders by contributing to the establishment of systems biology research on a European level via the European Science Foundation.

Finally, the boundary of collaboration is part of its scale. While the scope of VIRGO is limited through national boundaries, the Silicon Cell Initiative and the Census of Marine Life are much more open: boundaries are permeable and subject to change. If in the case of VIRGO, the project boundaries are predefined by the national funding organisations, in the very large international Census of Marine Life such a closed structure would not have worked. It would not have been possible to gather the different research endeavours with their different structures and sources of funding that are now part of the Census within a formalised, bureaucratic organisation with clear boundaries. The Census has open boundaries, which also explains why the project continues to grow and evolve, and why also scientists who are not formerly part of the project are invited to include their results in the database. In the case of the Silicon Cell, the boundaries are changing during the development of the project, in interaction with the search for collaborators and funding sources. The initiative started out on a national level with a branch in South Africa and now the systems biology proposals are being shaped in a European context. In short, network boundaries can be clear and static or permeable and changing. Where boundaries are not pre-set, networks may grow and evolve. However, at some point also very large-scale science needs to set boundaries and define its scope and the terrain covered.

When discussing the structure and size of different large-scale collaborations in biology, it becomes clear that the form of networks is often closely related to the character of research. More specifically, the analysis of different collaborations in biology shows how the structure of collaboration is related to the aim of research and how the size of collaboration can be shaped by research material or the objective of research. This means that collaborations combine a specific way of doing science with a particular organisational form in which research is performed. Following the definition of 'style' from art historian Ernst Gombrich (1968/72), I propose to use the notion of style to indicate this link between content and form in the context of scientific collaboration. By using 'style of collaboration' I build on various studies of science that already use the notion of style to indicate the intricate relationship between the research subject, the way of performing research and the social environment in which the research takes place (Fleck, 1981; Kwa, 2005; De Wilde, 2001). Although Pickstone (1993; 2000; 2007) does not explicitly use the word style, his 'ways of knowing' also intends to explicate these relationships.[171] In studying scientific collaboration, I would argue, styles of collaboration may specifically pertain to sorting things out, transforming information and applying knowledge.

The different styles of collaboration combine different orientations and deliverables with a specific organisational structure. First, the Census of Marine Life, given its aim to classify and catalogue all living animals in the oceans, can be characterised as a natural history collaboration geared towards sorting life out. The Census comprises a global project with a decentralised and participatory governance structure, enabling the participation of different research projects and the incorporation of as much research results as possible into the database. Second, the Silicon Cell project involves a practice of collaboration in laboratory biology concentrating on the transformation of information: analysis and experimentation. Building on the Human Genome Project – which focussed on the analysis of biological material – the Silicon Cell is dedicated to the building of a model on the basis of analysis and experimentation. Although laboratory biology does not have a long tradition of collaboration, the scientists of the Silicon Cell Initiative and the subsequent efforts in systems biology have the ambition to be big. They want to assemble different research groups in a semi-bureaucratic structure in order to standardise research efforts to build a common model and compress research time. Finally, the VIRGO consortium is

---

[171] See also Chapter 2: Collective ways of knowing, p. 53.

committed to the application of knowledge and collaboration with industry. As the creation of a shared product requires the different research groups to build on each other's work, this interdependency is facilitated by a bureaucratic structure and pre-defined boundaries (see Table 3).

	Sorting things out	Transforming information	Applying knowledge
Casus	Coml	Silicon Cell	Virgo
Deliverable	data	knowledge	product
Governance	participatory	semi-bureaucratic	bureaucratic
Interdependency	low	medium	high
Size	international	European	national
Border	open	in formation	closed

Table 3: an overview of different styles of collaboration in biology

Certainly, it should be noted that the different collaborations do not strictly reflect a single style. For instance, while the emphasis of the Census of Marine Life is in sorting out life in the oceans, the project also incorporates new developments in laboratory biology and is concerned with the application of knowledge. And next to the application of knowledge, the VIRGO consortium also focuses on transforming information. Thereby, some features of collaboration are more attached to a certain style of collaboration than others. For instance, deliverables play an important role in the determination of a style of collaboration, while size is more variable. In addition, the styles of collaboration are dynamic. Pickstone (1993; 2000; 2007) has shown how the various ways of knowing emerged in the course of history, starting with natural history after which the other styles came into being. Similar to the different ways of knowing, the various styles of collaboration may exist side by side, while they also interact and transform. This became visible, for instance, in the Census of Marine Life, which integrates traditional natural history collaboration and new developments in analytical and experimental laboratory biology into what I called 'new natural history collaboration'. As styles of collaboration change and develop, it is quite possible of course that new styles of collaboration will emerge in the future.

Rather than centring on the actual performance of research, collaboration in biology may also be geared primarily towards the construction of infrastructures. An example of an infrastructure project is the building of the Technology Facility at York University that I described in Chapter 2. Similarly, the Netherlands Genomics Initiative, which develops policies and orchestrates research, operates as an infrastructure project. These infrastructural collaborations can be viewed as a new style of collaboration in biology which is part of the development towards big biology. When examining these projects more closely, I found collaborations that have a resources orientation (both material resources and data resources); an economic orientation (projects to stimulate innovation); a policy orientation (projects that design policies or distribute funding); a regulatory orientation (projects responsible for the construction of regulation); and an interaction orientation (projects dedicated to the exchange of information and debate).

When taking a broad view on collaboration in biology, then, a wide array of projects may be identified. Although a distinction can be made between research projects and infrastructure projects, the orientation of collaboration ranges from making an inventory of life towards setting up biotech business or stimulating discussion about new developments in biology. These different orientations of the projects are tied to different deliverables the projects provide, such as data, knowledge, products, services and regulation (see Figure 2).

PROJECTS	RESEARCH			INFRASTRUCTURE
Objectives	Sorting things out	...Transforming Information	... Applying Knowledge	Orientation: Resource Policy, Economic, Regulatory, Interaction
Case studies	CoML	SiC	VIRGO	
Deliverables	Data	..... Knowledge	..... Products	Services Regulation

Figure 2: an overview of research and infrastructure projects in biology

Finally, it is important to recognise the connection between research collaboration and infrastructure projects. This connection can be seen in the construction of new infrastructures to facilitate research and in research projects where the development of infrastructure has become a key activity. This intertwining of research and infrastructure is especially characteristic of the supersizing of science.

Building scientific collaboration

The organisation of collaboration is the result of a social process in which collaboration is constructed. After looking into the organisation of scientific collaboration in biology, I will now consider how collaboration is built. If the three case studies represent different styles of research collaboration in biology, they also demonstrate different construction processes: the restructuring of traditional collaboration in the Census of Marine Life; the creation of a new structure for collaboration in the Silicon Cell; and the combination of different orderings in collaboration in the VIRGO consortium. Although the different styles of collaboration in biology require different types of connection work, they also show general features of the building of scientific collaboration: the supersizing of science. Next to the formation of collaboration, the building of scientific collaboration brings to the fore two significant developments that have not been explored so far: the increasing use of project management to organise research and the rise of a new type of scientist.

The formation of scientific collaboration

Based on my investigations in the building of collaboration in biology, I distinguish three phases in the process of forming collaboration, which I will analyse below: the origin of collaboration, the building of connections and keeping science big.

*The birth of scientific collaboration

When scientists talk about the formation of collaboration, they predominantly focus on the goals and the reasons for collaboration, but not on the intellectual or social conception of the project. When asking scientists about the birth of their project, they have difficulties telling a story which often comes in bits and pieces. In their view, the project evolved naturally from earlier work and is not seen as a turning point. Only when explicitly asked some scientists of the Census of Marine Life tell what Knorr-Cetina (1999) calls a birth-drama: a shared story of origin. In case of the Census the story goes as follows: the two initiators meet during their holidays at the seashore and decide to start the project. However, most of the scientists I talked to were even a bit surprised that the birth of collaboration would be of any interest. Nevertheless, they did enjoy going back in time to do some 'history of science'. One scientist even concluded to his own surprise that the project was very much interwoven with his own interests and scientific career, which was in contrast to how he thought it was: the project being just the next step in the linear trajectory of scientific progress.

The birth of collaboration in biology can generally be found in research, or it can be the result of policies, research policies or economic concern. In addition, inspiration for collaboration may come from public concern originating with NGO's, media or citizens, or from individuals sharing a common concern (e.g. a patient group that builds a research network around a rare disease from which its members suffer). If these different inspirations may well be intertwined within specific projects of course, the collaborations I have studied predominantly have a research inspiration. The Census of Marine Life builds on existing research into ocean life and aims to acquire insight into biodiversity in the oceans. The Silicon Cell emerged from a successful experiment, using a model of a specific process in a cell. And the VIRGO consortium is the result of the development of a line of research – combining virology and genomics – by the scientist who later becomes the project coordinator.

The birth of a project cannot simply be divined from the way the project looks when fully functional, as projects constantly transform during the building process. In the case of the Silicon Cell initiative, I have analysed in detail how the ideas for collaboration and the organization of collaboration constantly transform each other in the process of building collaboration. My other case studies confirm this interaction between the scientific and organizational aspects of collaboration. Based on their current research, scientists develop a new research approach and gradually translate this into a large-scale research project that is then also placed in the context of other concerns. Consequently, during the formation of collaboration ideas about collaboration are constantly influenced by other actors. However, the proportion of influence of various actors differs. In the case of the Silicon Cell and follow-up efforts in systems biology, the initiators kept very much in charge of the collaboration design and their original ideas. In contrast, within VIRGO the first ideas for collaboration transformed quite substantively through requirements linked to funding policy.

*The building of connections
In contrast to the birth of collaboration, scientists vividly recall the work that is involved in the building of connections. Within all the projects connections with scientists need to be built, as well as connections with the policy domain, industry or the public. What becomes especially apparent when studying the construction of various research networks is the difficulty of making connections. Although living in a so-called 'network society' (Castells, 1996), the construction of a network entails lots of connection work which is often hidden when talking in terms of networks (Barry, 2001). In other words, the making of connections is an art. Moreover, different styles of networks require different forms of connection-work.

Aiming to become an international collaboration, the Census of Marine Life builds international relations: networking with scientists, funding organisations and –influenced by the Sloan Foundation – also with the public. As a natural history style of collaboration with low interdependency between researchers, the connections do not have to be very strong and are therefore relatively easy to establish. Nevertheless, the project experiences difficulties in becoming truly global. Time is needed to get scientists involved and global inequalities make it difficult to expand collaborations worldwide. Because the Census has open boundaries, the building of connections involves a continuous process while connections are primarily built through communication and participation. Census members try hard to get the project and its goals known in the international ocean research community and involve scientists in discussions about the project. And scientists are asked to contribute their research results to the database. In addition, connections to the policy domain are established by means of extensive lobbying and drawing a connection between the research project and problems of biodiversity and climate change. In turn, the establishment of connections with the public calls for translation of research into popular media, such as exhibitions, newspaper articles or movies.

As a new type of collaboration in laboratory biology the Silicon Cell initiative especially showed what it takes for scientists to establish connections. The scientists are building a research network based on existing contacts with fellow scientists, but most importantly need to build connections with the policy sector in order to get funding. As a result of the policy emphasis on application, connections with industry are built as well. The building of connections takes time and asks for special skills, such as lobbying, the development of strategy and translational work, like the writing of research proposals and the making of a website. However, in this case it proves hard to reach concrete results because large-scale collaboration in laboratory biology is relatively new and the field of systems also has to be established. In addition, the consolidation of connections with policy and industry is experienced as difficult by initiating scientists. They explicitly do not want to become involved in negotiations full-time, as they want to protect some research time in order to keep their scientific career. As a result, connections with scientists, policy and industry have been built – materialising in some research collaboration – but connections did not yet solidify in the large-scale project envisioned.

Within the VIRGO consortium, connections are built on already existing contacts, in science as well as in industry, and solidified in an 'innovative cluster' within the Netherlands Genomics Initiative. This case study revealed how the different societal orderings of science, government and industry are brought together in the project format. Here relations are tight and contractual and their development is guided by funding requirements. Consequently, the work involved in establishing the collaboration is predominantly the moulding of research plans and participating organizations into the predefined funding format. This makes the building of connections in VIRGO different from the building of connections in other projects, as this is a form of 'brokered collaboration' (Shrum, et al., 2007), which means that funding agencies play an important role in bringing parties together. VIRGO is a clear example of a collaboration in which the funding policy shaped the building of connections, by setting the requirements and limits of connection building: international collaboration is not allowed, but connection with industry is required.

Next to the building of social connections, the construction of material connections is crucial in the formation of collaboration. The different case studies show the interweaving of social and technical networks around for instance technologies and data. I suggest to characterise these socio-technical networks as the cement of collaboration in biology. As collaborations have a networked character, standardisation plays an important role in the coordinating between research practices in the different research sites. Standardisation of sequencing technology has been an important part of the construction of the Human Genome Project (García-Sancho, 2008). In addition, the standardisation of data is crucial for the creation of databases. Databases are a crucial component in research networks and in many cases they are the backbone and lasting result of scientific collaboration (if ways are found to maintain the databases). They also form the basis for the modelling and visualisation of biological processes. Moreover, within biology an important reason for standardisation can be found in the living research material. For instance, within the VIRGO consortium they need to make animal models to perform research. As a result different types of standardisation can be distinguished: standardisation of technologies, standardisation of research material and standardisation of data. All three forms help to discipline and facilitate collaboration.

My case studies in biology also show how the making of big science requires extensive lobbying and creativity in the building of connections. Especially when it concerns a discipline in which collaboration is not part of normal practice or when it comes to more fundamental research without clear applications, it can be very difficult to construct bigness. This is a source of frustration for life scientists, especially when they compare biology research to big physics or space research that in their eyes acquire money relatively easily. "How do they do that?" (interview Van Driel, 2005). However, when talking to Prof Dr Kors Bos — a physicist working at the European Research Organisation for Particle Physics (CERN) who is involved in the coordination of the ATLAS experiment, one of the newest big physics projects — he suggests that physics research would never have acquired bigness in the current policy environment that focuses on innovation, and that physics still owes its bigness to the after-war period (interview Bos, 2008). This also nicely points out that although the building of connections can be difficult, once connections have been built, they prove to be fairly stable. This resilience of big science can be seen in physics and space research (Lambright, 1998), but also in biology. The Netherlands Genomics Initiative has already managed to extend its own life and also the Census of Marine Life is looking for ways to prolong the collaboration after its initial ten years of funding from the Sloan Foundation.

*Keeping science big
Just like every form of science, scientific collaborations claim to push the frontier of science, to make the invisible visible or the unknown known. However, in the case of large-scale collaborations it is claimed that big steps can be taken at once: big steps on the road towards ultimate knowledge. In other words, the collaborations are building an image of bigness. In the stories of scientists visions of completeness are central to legitimate large-scale collaboration. The supersizing of science comes together with great romantic visions of science resembling the encyclopaedic ambitions from the 19th century. This big thinking is reflected in the project narratives, which sell the collaborations to science, politics and the public. For instance, the Census of Marine Life claims to count *all* the animals living in the oceans, constituting a first important step towards understanding the historic and future dynamics of ocean life. Similarly, the Human Genome Project was set up to make the *complete* map of the human genome. Knowing the building blocks of human life was presented as a first step towards understanding health and disease. This staging of big science already takes place during the building of collaboration, but continues after connections are established in order to keep it big.

When looking further into the legitimising of large-scale research projects, space and time turn out to play a critical role, as the collaborations want to compress time and open up space. Space is important as the subject of research is very small (as in genomics) or, conversely, very large (as in space research), and complicated technologies are vital to get access to these spaces. In the Silicon Cell and VIRGO projects, large efforts are needed to work at the small molecular level – generating a mass of data which occupies a huge 'digital space', but in the Census of Marine Life the spaciousness of the oceans requires a large-scale approach precisely to make the data more compact and workable. In addition to covering the two extremities of space, the large-scale research projects want to compress the time that is necessary to perform the research by collaboration. They quicken the 'normal' pace of research through collaboration, which is claimed to be highly beneficial for the progress of research. In addition, a sense of urgency is created: now is exactly the right time for the research effort to take place. It could not have been done earlier, but by not doing it now and within a reasonable timeframe, precious time is lost and scientific progress obstructed (which will come at a high cost: people unnecessarily dying or environmental damage). Finally, the research collaborations want to intervene in time. The projects claim a changing effect on the future: a better understanding of life also makes it more 'manageable'. For instance the Silicon Cell wants to find a cure for cancer and VIRGO will help to prevent future pandemics.

Nevertheless, the huge scientific and societal expectations that are created in large-scale collaboration often lead to disappointment afterwards.[172] The Human Genome Project is a clear example of a project that did not live up to its promises: the 'book of life' proved not readable after all and turned into a pile of data, transforming from meaningful to unmeaningful information. Now, research projects are concentrating on the interpretation of these data. Systems biology, for instance, promises to make sense of the analytical data via modelling. However, it is unclear yet how the promises of systems biology will be evaluated and – as became clear from my case study on the Silicon Cell – there is already discussion about its potential. Moreover, within VIRGO the promise to develop a new therapy or vaccination has been put into perspective already: now the aim is to develop screening technologies, which seems a much less heroic objective (Boekholt et al., 2007). And finally also the Census of Marine Life will not be able to deliver its promise to catalogue all animal life in the

---

[172] See for expectations in science and expectations in scientific collaboration, respectively Brown et.al., 2000 and Douglas, 2005.

oceans in the near future. In sum, big ambitions are materialised in practice only partially at best.

This touches on the fact that big science often appears bigger than it actually is. Importantly, the process of supersizing science tends to hide fragmentation and uncertainties. This becomes visible in expectations that are not met, but also in the absence of connections. After hearing about the ambitions of the Silicon Cell initiative, who would think that only a couple of people are performing the research? And the innovative cluster that successfully linked academia with industry was basically linking one department to its own spin-off. Moreover, one of the reasons that the Census of Marine Life will not reach its goal is the impossibility of getting access to all parts of the oceans, as not all nations participate. In addition to hiding fragmentation, big science does not leave room for reflection or uncertainty. As projects need to accomplish things within a relatively short amount of time, they leave no room for hesitation or reflection. This became apparent in the set-up of the Netherlands Genomics Initiative, whereby the hurriedness left no time for a careful weighing of all the options. It is not clear what the long-term effects of this masking of uncertainty will be, especially if uncertainty around, for instance, issues of standardisation is involved.[173]

The projectification of science

An important difference between the work of Watson and Crick on the double-helix and contemporary research collaborations is the project format in which research and especially scientific collaboration take place. With important roots in the Manhattan Project and the Apollo space programs, project management developed in fields of construction and engineering during the 1960s (Cicmil & Hodgson, 2006; Hodgson, 2004; Lock, 2003). The 1990s saw the project mode expanding across industries and other sectors, which is aptly described as the 'projectification of society' (Midler, 1995). Science is also increasingly seen as a manageable process, featuring strategies, roadmaps, and increasingly takes shape as a 'project' with its own acronym and a special logo.[174] However, reflection on the meaning and impact of the project in scientific practice cannot be found – neither in studies that critically reflect on the universal adoption of

---

[173] For an analysis of uncertainty around the standardisation of stem cells in the International Stem Cell Initiative, see Eriksson, 2008; Webster & Eriksson, 2008.

[174] It should be noted that projectification not only plays a role in scientific collaboration; frequently, smaller science also takes place in a project format. For instance, it is quite common to talk about a 'PhD project' in which only one researcher is involved.

the project as a way to organise society, nor in studies on scientific collaboration or scientific work. Based on the projects I analysed, I explore what I call the 'projectification of science'. After a short introduction of the 'project' format and 'project management', I will argue that projects are a way of packaging inquiry more formally, through a design that considers a clearly defined problem that has a solution and a deliverable at the end. The discourse of 'the project' acts to mark out a specific time and space horizon within which the project is to be undertaken. Moreover, it also implies that those doing the work must be prepared to be evaluated.

In general, the propagation of the project mode of management is accompanied by a discourse on the project as an organisational response to the challenges of managing in a world of growing complexity (Cicmil & Hodgson, 2006; Hodgson, 2004; Lock, 2003; Midler, 1995; Sahlin-Anderson, 2002). The projects' origin in modernity gives it a rational basis and a functionalist and instrumental view, focussing on time, cost and output. But at the same time the project is presented and adopted as a new working mode in late- or post-modern societies that replaces the modern, hierarchical bureaucratic mode of organisation. Projects promise to deliver the ideal combination of a versatile and flexible but predictable form of work organisation, one that delivers controllability and adventure as well as decentralisation and accountability. However, more critical accounts on project management show that the world of projects is inhabited by dilemmas and contradictory logics. In contrast to what the project rhetoric suggests, project work often does not work. There seems to be a gap between project theory and the use of project management: "contemporary studies of project performance continue to indicate the disparity between the maturing body of project management know-how and the effectiveness of its application" (Cicmil & Hodgson, 2006: 6). Also, it is common knowledge that many projects cost more time and money than projected. Evaluation reports and studies provide insights in frequent cost overruns, substantial delays and under-performance. So while the project is widely adopted as a research format, its effectiveness as a management strategy is subject to discussion.

What does the adoption of a project mode of working in scientific practice entail? Most obviously, the scientific project formalises scientific enquiry, via diverse forms of contracts; legal, financial and technical. In addition, the project enables the combination of different societal orderings, as shown in the context of VIRGO. The project format functions as a boundary technology in which the different interests and approaches of science, government and industry can become aligned. Importantly, this alignment takes place through the creation of the project as a separate organisational entity, with its own name, acronym, logo, website and project stories. Each project comes with the creation of a narrative constructed by people talking and writing about the project (project descriptions and proposals). This narrative legitimises the project and contextualises it, by embedding it into broader narratives like discourses on scientific progress or societal problems. However, these stories obstruct discontinuities or disagreements, as well as the work behind the project scenes. Project goals and organisation (input) and results as data or publications (output) are presented, but the 'scientific work' is hidden behind the official project face.

However, projectification changes scientific practice more profoundly. First of all, the project mode of working has added an extra phase to scientific practice: the construction of a project proposal has become the first step in doing research. This means that research now has to start with a clear-cut problem and concrete objectives in mind, which does not leave space for gradual scientific development or surprise and also encourages the making of scientific promises. Moreover, deliverables often differ from traditional scientific deliverables (data, publications) but also need to show the relevance and applicability of the scientific research. In addition, the creation of a project also means the creation of a separate time and space for science. Dispersed scientific locations become connected in the project structure, and projectification transforms the pace of science by marking out a specific time period in which research has to be done. As the etymology of 'project' suggests (Latin 'projectum' from 'projicere', which means 'to throw something forwards'), time is an important factor in projects. This indeed resonates with the big science rhetoric of doing science faster by creating a large-scale project. The project creates its own internal clock; its own schedule with deadlines. As a result, scientific time becomes compressed and can be more easily controlled (Adam, 2004).

This control of time aligns with the emphasis on accountability and evaluation in project management. Large investments in science encourage the concentration of control in science (Whitley, 2000) and the project format makes science more open to external control. According to Hodgson, project work involves hybrid modes of control that interweave two broad logics of control: "the autonomous, self-directed ethos of the post-bureaucratic subject and the bureaucratic exigencies of structured work and systematic reporting procedures" (Hodgson, 2004: 98). For science this means that next to traditional forms of scientific evaluation – like peer-review – projectification makes legitimisation and accountability a key activity, not only at the beginning and the end of projects, but also at regular intervals in between. Moreover, standards for evaluation have become more diverse, not only focussing on scientific results but also on process and societal standards. The projectification of science thus makes the messy research process more transparent and better manageable for other actors than scientists. Consequently, the projectification of science and society at large is linked to the increasing audit of both, which has already been captured in the idea of 'audit society' (Power, 1997).

Nevertheless, a scientific project has a different meaning for different participants or parties involved. From the point of view of scientists, projects are a temporary part of a scientific career. The project is often just one of several activities that are created around the central line of development of their personal scientific interest, embedded in the international scientific community and its history of progress. When taking a policy point of view, projects are seen as individual units, which are part of a specific family of projects with a specific aim, be it the development of a general research field or the stimulation of innovation within a specific country or region. And for the analyst of scientific collaboration, the project can be a research entity. So the same project has a different function for different actors, leading to a different embedding of the project in different contexts. As a result, the project takes a position in various landscapes: in the life sciences field, in the big science world, in the policy environment and in society at large. Next to institutional positioning, different other forms of positioning take place: financial positioning (where is the money coming from), technological positioning (what kind of technologies are used or developed) and spatial positioning (involving different aspects from buildings to geographical positioning and ways in which connections are formed).

The ways in which projects position themselves and are positioned vary, and are closely related to issues of legitimisation. For instance, when – as in the case of the Silicon Cell – a project is the initiative of scientists, scientists need to legitimise it towards the policy sector and the public. But when a project is brokered by policy, the project has to be legitimised vis-à-vis scientists. In addition, criticism plays a role in the positioning of projects. Criticism can tackle the scientific approach, the organizational part of the collaboration or both. The Human Genome Project was scrutinised for its reductionism as well as its industrial approach and systems biology received criticism as being the emperor's new clothes and just another name for something that has been done for ages. Confronted with criticism a project needs to position, by countering the arguments or by emphasizing positive aspects of the collaboration. Criticism seems to matter most when the project is in its starting phase and still trying to legitimise itself. When already up-and-running the importance of critique diminishes as it does not influence or endanger the project anymore.

Transforming roles of scientists

The general discourse on big science already indicates transforming roles of scientists in large-scale collaboration. Most importantly big science is taken to imply a loss of creativity and autonomy in the large industrialised and bureau-cratic scientific structures. Scientists are envisioned to become line workers in the large-scale factory of science. Although in the context of the Human Genome Project indeed boring repeating tasks are reported – mostly performed by PhD students and technicians – my research shows other transformations in the role of scientists in large-scale research too. Interestingly, I found that individual scientists play an important role in the collectivization of research. The case studies display engaged scientists who put a lot of work in the formation of collaboration and are the driving force behind it. Also in collective research, then, the individual scientist fulfils a crucial role. However, I also showed how the roles of these scientists are transforming in the context of collaboration. More specifically, my case studies show the creation of new roles for scientists, the creation of hybrid scientists and also the disappearance of some traditional roles of scientists.

First of all, the creation of new roles became apparent in relation to technology development and the construction and management of large-scale research. With the development of technologies in biology and the emergence of new socio-technical networks, the importance of technicians and bioinformaticians increases. In addition, the case of the Silicon Cell clearly shows how in the setting-up of collaboration scientists need to perform new roles that ask for special skills. The building of social and technical connections is a key activity in the formation of collaboration and the building of connections is an art. Although being a scientist always involved the translation of findings towards a broader (scientific) public and the embedding of specific research in a network of knowledge (Latour, 1987), networking becomes more important in the context of scientific collaboration and also crosses the border of the own disciplinary or scientific community. Science becomes more clearly a form of lobbying and negotiating with other scientists, policy and industry. In addition, public communication becomes more important, which could be clearly seen in the Census of Marine Life. Finally, large-scale collaboration brings new roles in research management, like project leader and project coordinator. As a result, scientists need different skills and new types of scientists emerge.

The creation of new roles in science causes tensions, on the individual level of the scientists, as well as in the science system as a whole. Although, the case studies showed how content and organisation of research are mutually shaping each other, individual scientists make a clear distinction between 'doing science' and 'organizing science'. They generally present research as the important part of their work and the organisation of research as a necessary evil. While some scientists like to present themselves in various different roles – as the hybrid scientists discussed in the context of VIRGO – other scientists have difficulties combining the various roles that need to be performed in collaboration. The networking that is necessary to set-up collaboration involves politics and strategy and some scientists claim not to be trained for this. Or they simply do not want to invest too much of their time in the organisation of science. Also, the lobbying and negotiation that is involved in setting-up a new collaboration can be seen as antithetic to traditional scientific values, like disinterestedness and objectivity. Moreover, the new roles challenge the scientific system, as they are generally not part of traditional career development and the current research system often does not accommodate these new roles. For instance it does not provide standard funding for technicians, and research management is not a recognised career in science. In the opposite direction, it is also increasingly difficult to get funding for the traditional role of taxonomist, as became visible in the Census of Marine Life.

In turn, the development of new roles in science sheds a new light on the general ideas of loss of creativity and autonomy of scientists in big science resources. Although sequencing or other semi-automated large-scale research may be less creative than more experimental or theoretical forms of research, I suggest that large-scale science does not mean the end of creativity. First of all, creativity may as well be located in the process as in the product. While in the case sequencing the process may be uncreative – although lots of ingenuity was put in technological developments – the product is something unique. Creativity can be found in the establishment and coordination of large-scale collaboration. Also scientists are inventive in opening up possibilities to get research funding and adapt the appearance of research and its organisation to funding requirements. Moreover, it is argued that one of the benefits of automation is that it allows the 'big science' aspects to be handled elsewhere, while the real small-scale and epistemologically interesting work could be done in the lab.

Still, the decrease of autonomy is indeed a price scientists have to pay for collaboration, as big science's pre-set problems, clear goals and need for societal relevance and accountability do take away some of the autonomy of scientists. However, sometimes it is worth to pay this price. For instance when collaboration brings something that could not be reached alone. Or when wanting to develop practical knowledge, societal embedding is better suited than the ivory tower. In addition, it is argued that bureaucracy not only decreases the autonomy of scientists, but can also be seen as a way to create a separate space for science (Shrum et al., 2007). In addition, it may well be that big science opens up new space within it for small scale science that pursues matters in a more open way. Finally, the relation between autonomy and intellectual property plays an important role. When it comes to the development of potentially important new ideas, scientists will be less eager to collaborate, or only within a small group.

The many faces of collaboration

Scientific collaboration is a multi-faceted phenomenon. As I have shown in the case of biology, there are many perspectives on collaboration. I started this analysis with the different normative views on the increase of collaboration: it can be seen as a positive as well as a negative development. When taking an empirical perspective, many reasons can be found for forming collaboration. I have shown how the expansion of biology is influenced by the interplay between scientific, technological and social developments. In addition, I presented different styles of collaboration in biology: sorting out data, transforming information and applying knowledge. Despite their differences, the case studies do show similar themes and features, as for instance the use of information and communication technologies, the projectification of science, the push towards application of knowledge and changing roles for scientists. These are characteristics that also come into play in collaborations within other scientific fields. Consequently, the increasing collaboration in biology can also be seen as an exponent of a broader trend towards collaboration in various scientific fields, which makes my analysis relevant for scientific collaboration in general.

Most importantly, the case of big biology informs about the building of scientific collaboration in our present-day society. In the building of collaboration, different elements play a role. On the one hand social factors are important, like the size, the management structure and the influence of scientists, funding organisations, science policy and politics. On the other hand, material factors determine collaboration. For instance, the research material, the technologies used and the type of research output can have an influence on the organisation of collaboration. These social and material aspects of organisation can intertwine, as for example in organisational movements around technologies and processes of standardisation. In addition, I distinguished different processes of building collaboration that exposed different types of connection work: the restructuring of traditional forms of collaboration (as in the Census of Marine Life), the creation of a new structure for collaboration (Silicon Cell) and the combination of different orderings in collaboration (VIRGO).

In analysing the case studies I showed novel perspectives on scientific collaboration as a multi-faceted phenomenon. First, I distinguished different styles of collaboration and their dynamics. The Census of Marine Life illustrated how different styles of collaboration merge. The traditional natural history collaboration mixes with new styles of collaboration in molecular biology. This becomes especially apparent in transforming practices in taxonomy. Secondly, the building of the Silicon Cell showed how scientific collaboration has a front and a backstage. There is a difference between the staging of collaboration to diverse audiences and the actual collaborative research that is taking place behind the scenes. The Silicon Cell illustrated how the presentation of collaboration is a crucial part of the building of collaboration, but this case also showed how front stage and backstage can diverge. Finally, the analysis of the VIRGO consortium explicated how collaboration has a different face on the level of management, research practice and the individual level of the scientist. Although science, industry and government merge within the VIRGO project, this does not mean that collaboration takes place at all levels. Boundaries between the different domains also become realigned or are kept steady. This exemplifies how within collaboration, fragmentation can be found as well.

As Galison and Hevly (1992) already stated: big science has many faces. It cannot only be studied from different perspectives, the concept also informs the historical development of large-scale collaboration in general and the different disciplinary forms of collaboration. By studying the increase of collaboration in biology from a big science perspective, large-scale biology research becomes part of the big science family, which also includes physics, astronomy and space research. I showed how field biology already had a tradition of collaboration, while laboratory biology collaboration only recently has become important, in interaction with scientific developments on the molecular level and related technological developments. In addition, I showed how big biology is a networked form of big science that contrasts with the centralised structure of big physics. Biology is characterised as a contemporary form of networked large-scale research collaboration, which can be again divided into different styles of collaboration. However, this networked form of big science is not restricted to the biological realm as its characteristics can be identified in various other forms of current scientific collaboration.

Although some characteristic of biological research are disciplinary characteristics – like for instance the research material being alive – the analysis of big biology can be useful for understanding other forms of contemporary big science. This contention is confirmed by contemporary developments in big physics and nanotechnology. More recent studies of big physics show how the centralised and industrial modes of production that resulted from World War II are now transforming into a network direction too (interview Bos, 2008; Galison, 1997; Galison & Jones, 1999; Van der Heijden, 2008). The 1970s brought a new era of dispersed production with peripheral sites and a decentred subject, which also developed in interaction with the development and use of information technologies. Currently, the so-called 'grid' or 'web 2.0' is developing in the context of physics research to deal with the enormous amount of data that are generated in new experiments.[175] In addition, developments in nanotechnology – that is often envisioned as the next big thing – are accompanied by research programs that look into societal implications.[176] This clearly follows the example of ethical, legal and social research in big biology. As my characterisation of big biology shows how science does not necessarily need large centralising instruments to become big, it can help to explore the expansion of forms of science that lack these centralising technologies.

In addition, my analysis of big biology can form a starting point for exploring the increasing scales of research beyond the natural sciences as large-scale networks are increasingly proliferating in the social sciences and humanities as well.[177] For instance, the integration of information technologies stimulates large-scale research projects in sociology, as exemplified by a project that uses internet to explore how blockbusters are made (Hedström, 2006; Salganik et al., 2006). While an example of big history can be found in the large-scale research project 'Tensions of Europe', which explores and defines ways to study transnational European history.[178] Also in Science and Technology Studies, large-scale projects become more common, like for example the PRIME network which concentrates on science policy and innovation.[179] This network is a so-called 'network of excellence', one of the instruments with which the formation of international research networks within the European Union is stimulated. Finally, ethical, legal and social research connected to biology often takes place in networks, like the Genomics Network in the United Kingdom.[180] How do large-scale research networks in the social sciences and humanities relate to other forms of big science? Studies that try to answer this question

---

[175] In the context of the development of web 2.0 and its uses beyond physics and the natural sciences, computer scientist Schneidermann suggests the development of 'Science 2.0': collaboration-centered socio-technical systems in which research is performed (Schneidermann, 2008).

[176] See for example the Center on Nanotechnology and Society and Nanoned TechnologyAssesment. Retrieved February 28, 2008 from http://www.nano-and-society.org and http://www.nanoned.nl/TA/

[177] An example of research that looks especially into research networks in the social sciences and humanities is the research project of Dormans & Kok from the Virtual Knowledge Studio of the Royal Netherlands Academy for Sciences (Retrieved September 7, 2008 from http://www.virtualknowledgestudio.nl/projects/socio.php).

[178] Retrieved November 2, 2008 from http://www.tensionsofeurope.eu

[179] Retrieved February 28, 2008 from http://www.prime-noe.org

[180] This network exists of 4 research centres located at universities across the UK. Retrieved February 28, 2008 from http://www.innogen.ac.uk/About/ESRC-Genomics-Network.

may well contribute to the expansion of knowledge on the supersizing of science and the contemporary big science family that now also includes biology.

EPILOGUE

The future of science
The next generation

During my study into large-scale research networks in biology, I encountered another type of network in which I became personally involved and which supported me in writing this thesis: networks for young life scientists. Although these networks are not primarily dedicated to research, they should be mentioned in a study on collaboration in biology as another type of network that recently emerged in biology. As I experienced these young networks are flourishing and soon after starting my research I became involved in no less than five different groups. Initially, I became a member of the Genomics Network for Young Scientists (GeNeYouS) that aims to bring together Dutch PhD students, post-docs and other young scientists who work on genomics and related research projects. I was one of the first social scientists who joined the network, but later I also discovered a network especially for social scientists working on genomics. Corsage is short for Cooperative Researchers on Society and Genomics and was founded in 2003 by a small number of PhD students from various universities in the Netherlands. About a year later Corsage merged with GeNeYouS and began collaborating with a similar social science network in the United Kingdom: the Post Graduate Forum on Genetics and Society (PFGS) of which I became a member during my Marie Curie Fellowship at York University. Finally, I became part of two international networks of young scientists. I participated in the BioVisionNxt. Class of 2005 that gathered 100 PhDs, Postdocs and MBAs from around the world. This was organised alongside the World Life Sciences Forum in Lyon with support of the European Commission and the Young European Biotech Network (YEBN), a network that wants to strengthen collaboration of European young life scientists that I joined as well.[181]

From this list of initiatives it becomes clear that young researchers have started to associate in various ways. When looking into the history of these networks, it seems that the first networks of young life scientists emerged in the last decades of the 20th century to support the somewhat controversial developments in biotechnology. In the more recent context of developments in genomics and post-genomics research and increasing networking in the life sciences, networks of young scientists seem to multiply as well. Network formation takes place in two ways: bottom-up versus top-down. Most of the networks are formed bottom-up. For instance, the Postgraduate Forum on Genetics and Society (PFGS) was set up by a couple of young researchers in 1998. One of its founding fathers, Richard Tutton, recalls how the PFGS had a humble beginning, with a first meeting in the senior common room of Boland College at Lancaster University on a wind-swept day in April:

> We had a budget of less than five hundred pounds, had to pour our own tea and coffee and illustrated our talks with the aid of a single ageing OHP. At that first meeting, it was agreed that another should be held under the title of Postgraduate Forum on Genetics and Society at University College London in December of that same year. Thereafter, the colloquia became an annual feature and were held across the country at various institutions. (…) From the first meeting which attracted ten postgrads, a greater number attended each meeting until at the Cambridge meeting there were 29 participants, with an increasing number coming from outside of the UK.[182]

In contrast, GeNeYouS was an idea of the Netherlands Genomics Initiative. Policy officer Luc Rietveld explains:

> Within the NGI strategies predominantly aimed at more established researchers that already have a track-record. And of course we funded PhD's and postdocs, but nothing was done to network young researchers, to organise the young-researchers field. That is when the 'Godfathers' Janneke Timmermans and me started to approach the young scientists. (interview Rietveld, 2006)

---

[181] These different networks present themselves on a website, retrieved November 9, 2008 from: http://www.geneyous.nl; http://www.geneyous.nl/corsage; http://www.pfgs.org; http://biovision.blogs.com; http://www.yebn.org.

[182] This is part of the story told by the cofounder of PFGS, Dr Richard Tutton, on the beginnings of the network, during the annual PFGS meeting in York on August 30, 2006. Tutton is now a senior lecturer at the ESRC Center for Economic and Social Aspects of Genomics (CESAGen), University of Lancaster, UK.

They invited the young scientists to a meeting where the possibilities for a network were discussed and foundations were laid. "We explicitly did not want to build the network: that was their job" (idem). Consequently, the network was built in 2003 by a group of enthusiast young scientists, who formed a board and set up different working groups with substantial (financial) support from the NGI.

When analysing the aims of the various networks quite some similarities can be noted. The main goal of the networks is interaction. In the network programmes this is articulated in a variety of concepts: networking, information flow, exchange of ideas, exchange of knowledge, multi-disciplinary cooperation, socializing, building a community and internationalisation. Different means are employed to reach these interactions. First of all, the various networks have their own website or online forum. In addition, they issue a newsletter and organize meetings, varying from annual meetings to regional workshops or presentation at major conferences in the field. Education and training are also important in these networks. This takes place within the meetings, for instance during a workshop networking or in a competition to develop the best research strategy for a pharmaceutical company. Likewise, separate courses are organized. For example, GeNeYouS has coordinated training on various genomics technologies and organized a visit to 'Saragene' were we got acquainted with the use of 3D technologies in genomics research.[183] The networks prepare for a future career, not only via networking and training but also through CV building, the organization and support of job markets or career fairs, as well as visits to companies or research institutes. Finally, communication with the public is on the agenda, which materialised for instance in the organisation of 'Biopop', a project in which young European biotechnologists meet citizens to build a new model of communication with society, moving towards public participation, discussion and decision sharing in science.[184]

Gradually I came to realise that these networks are a unique feature of the present world of biology and are an interesting source of information on how science is transforming, how careers of the next generation of scientists are shaped and how the future of science will look. The networks themselves also claim that they are influencing future scientists, or as YEBN says: 'We build the BioLeaders of tomorrow!' Within science and technology studies, this young generation of scientists is frequently not taken into account. Most famous is the study of Sharon Traweek, who analysed the career of scientists in big physics research in her book *Beamtimes and Lifetimes: The World of High Energy Physicists*

---

[183] Retrieved November 9, 2008 from http://www.sara.nl/projects/projects_07_03_ned.html
[184] Retrieved November 9, 2008 from http://www.biopop-eu.org.

(1988). She shows how the career of young scientists in high energy physics can be divided into three stages (undergraduate training, graduate training and research associate), with each stage being marked by distinctive intellectual qualities and the cultivation of a certain emotional state. With respect to graduate students she shows how they slowly become part of the scientific community, through the telling of stories of the generation in power and stories of success and failure. In addition, they learn to be meticulous and hardworking and the emotional state can be best described as 'afraid of losing their chance at success by losing time'. When taking this perspective for looking at the current generation of young life scientists, I observe that the young networks are an addition to traditional forms of education and support the ability of young scientists to interact and cross borders – skills that are becoming increasingly important in the present life sciences.

When asking young scientists what they have learned from their participation in the various networks, interaction or networking is mentioned most. One of the founding scientists of GeNeYouS, Maurice Zeegers who has already become a professor 5 years after finishing his PhD, states:

> When I look at my career, I think that networking has been essential, it is very important. (...) Networking is an investment in the future. (...) Interaction is the most important goal of Geneyous. It helped me to build connections with young scientists as well as with people from science policy. (interview Zeegers, 2006)

This is in line with the aim of GeNeYouS as imagined by the Netherlands Genomics Institute:

> Interaction is crucial. Young researchers are primarily busy with their own research – studying that one gene or protein – and it is important that they also look beyond that. To get in touch with people who do the same kind of research and who use the same techniques and run into the same kind of problems, so you can communicate with each other. (...) And networking is also important for their future career. (interview Rietveld, 2006)

The founder of PFGS agrees that its primary function was to build networks amongst members of the new generations of social scientists working on genetics:

> And, this, I think, turned out to be one of its greatest successes – it was through the PFGS that personal friendships were established and sustained.

And many of the people I met during my time as a member of the PFGS I still consider to be friends today.[185]

Together, the PFGS board shared the experiences of writing funding proposals for the colloquia to be held, organising schedules of talks and publishing articles.

Boundary crossing is also supported by the networks of young scientists. Next to the crossing of national borders – via meetings of the European network or the global mix of young scientists that gathered in Lyon – the boundaries between diverse scientific and societal domains are crossed. Within the networks, different specializations in the life sciences meet. In addition, the natural sciences and the social sciences and humanities come together: when GeNeYouS and Corsage merged, and when the PFGS got its own scientist specialising in cancer research. Furthermore, the networks of young scientists bring together the world of academia, government and business, by gathering young scientists with different backgrounds and perspectives on their future career, and by explicitly showing the various possibilities in science, business and policy. For instance, Christine Bunthof who was part of the first board of GeNeYouS is now working for the Netherlands Genomics Initiative as a coordinator of a European research collaboration in plant genomics:

> I have learned a lot from my experiences on the board of GeNeYouS. Especially, organisational and management skills. (...) I would probably have done this work also without my experiences with GeNeYouS, but GeNeYouS has given me the skills to feel at home in this job. Therefore, GeNeYouS has certainly given me a big advantage. (interview Bunthof, 2006)

In addition, various young scientists keep an eye open towards their possibilities in the private sector and some already actively pursue a career in business. A unique example of various forms of boundary crossing is Ashwin Sivakumar whom I met during BioVision.Nxt. He was raised and educated in India, did his master in bioinformatics in the UK, pursued a PhD in Finland and showed me three business cards: presenting himself respectively as a PhD candidate, the owner of the company Jugular and the founder of an NGO: the Bioinformatics Society of India. In short, the career paths and future plans of network mem-

---

[185] This is part of the story told by the cofounder of PFGS Dr Richard Tutton on the beginnings of the network, during the annual PFGS meeting in York on August 30, 2006.

bers illustrate how young scientists are exploring boundaries between disciplines or between academia, government, business and the public.

Finally, I am pleased to tell that during the final phase of my PhD research I became involved in yet another network for young scientists. The Marie Curie fellows that took part in the *New Genetics/New Society? Integrating Science, Society and Policy* program at the Science and Technology Studies Unit of York University decided to gather together and form a more permanent network for future research collaboration. This makes the number of networks in which I participate amount to six. It goes without saying that these networks of young scientists have offered me a wealth of information on the life sciences in general, the practice of research, the functioning of networks and the position of young scientists and I hope to build upon this knowledge in the future. But above all I enjoyed meeting these talented and inspiring young scientists and I want to thank Aaro, Andrea, Andrew, Antoinette, Ashwin, Bart, Bettina, Christine, Conor, Diego, Erik, Fleur, Govert, Gyula, Helen, Ine, Ingrid, Jamie, Janus, Julia, Kadri, Klaus, Luis, Mark, Martin, Maurice, Melita, Miguel, Nete, Sakari, Ragna, Rens, Richard, Tim, Tom, Tora, Wietse, Wouter and all the other young (social) scientists for sharing their stories. I wish you all a lively career and a wonderful life!

Bibliography

Adam, B. (2004). *Time*. Cambridge [etc.]: Polity.

Alberghina, L., & Westerhoff, H. V. (2005). *Systems biology: definitions and perspectives*. Berlin [etc.]: Springer.

Alberghina, L., & Westerhoff, H. W. (2002). *The Yeast Silicon Cell: a molecular systems biology approach. Information Memorandum for an Expression of Interest for an Integrated Project*. Retrieved August 29, 2008 from http://www.siliconcell.net/ysic/eoimemo.html.

Appleby, P. H. (1945). *Big democracy*. New York: Knopf.

Ausubel, J. H. (1997). *The Census of the Fishes: Concept Paper*. Retrieved September 21, 2006 from http://phe.rockefeller.edu/COML_concept/

Ausubel, J. H. (1999a). The Census of the Fishes: An Update. Retrieved September 21, 2006 from http://phe.rockefeller.edu/COML_feb1999

Ausubel, J. H. (1999b). Toward a Census of Marine Life. *Oceanography, 12*(3), 4-5.

Ausubel, J. H. (2001). The Census of Marine Life: progress and prospects. *Fisheries, 26*(7), 33-36.

Baker-Masson, P. (2000). *Federal agencies and private foundation collaborate to support ocean science and research*. Retrieved September 21, 2006 from http://www.coreocean.org/Dev2Go.web?id-=232848

Baker, M. (2004). Collaborative projects brewing one year later at Clark Center. *Stanford Report*, November 3, 2004. Retrieved February 14, 2007 from http://news-service.stanford.edu/news/2004/november3/clark-1103.html

Bal, R. A., Bijker, W. E., & Hendriks, R. P. J. (2002). *Paradox van wetenschappelijk gezag: over de maatschappelijke invloed van adviezen van de Gezondheidsraad*. Den Haag: Gezondheidsraad.

Ballard, R. D., & Hively, W. (2000). *The eternal darkness: a personal history of deep-sea exploration*. Princeton, N.J.: Princeton University Press.

Balmer, B. (1996a). Managing Mapping in the Human Genome Project. *Social studies of science, 26*(3), 531-574.

Balmer, B. (1996b). The political cartography of the human genome project. *Perspectives on science, 4*(3), 249-282.

Barker, R. G. & Gump, P.V. (1964). *Big school, small school* Stanford, CA: Stanford University Press.

Barnes, B., & Bloor, D. (1982). Relativism, Rationalism and the Sociology of Knowledge. In W. Hollis & S. Lukes (Eds.), *Rationality and Relativism* (pp. 21–47). Oxford: Blackwell.

Barry, A. (2001). *Political machines: governing a technological society*. London [etc.]: Athlone Press.

Beaulieu, A. (2004). From brainbank to database: the informational turn in the study of the brain. *Studies in history and philosophy of biological and biomedical sciences, 35*(2), 367-390.

Bell, D. (1973). *The coming of post-industrial society: a venture in social forecasting*. New York: Basic Books.

Berkhout, G. (2002). *Het Nederlands innovatiebeleid: tijd voor vernieuwing?* Den Haag: Ministerie van Economische Zaken.

Berman, M. (1983). *All that is solid melts into air: the experience of modernity*. London [etc]: Verso.

Bie, R. de (2004). *Scylla en Charybdis: laveren tussen interne en exteren onderzoekspraktijken*. Afscheidsrede Universiteit Utrecht.

Bijker, W. E. (2001). Understanding Technological Culture through a Constructivist View of Science, Technology and Society. In S. M. Cutcliffe & C. Mitcham (Ed.), *Visions of STS. Counterpoints in Science, Technology and Society Studies* (pp. 19-34). Albany, NY: SUNY Press.

Bijker, W. E. (2006). The Vulnerability of Technological Culture. In H. Nowotny (Ed.), *Cultures of Technology and the Quest for Innovation* (pp. 52-69). New York: Berghahn Books.

Bijker, W. E., & Hughes, T. P. (1987). *The social construction of technological systems: new directions in the sociology and history of technology*. Cambridge, Mass etc.: MIT Press.

Bijker, W. E., & Law, J. (1992). *Shaping technology/building society: studies in sociotechnical change*. London: MIT Press.

Blume, S. S. (1992). *Insight and industry: on the dynamics of technological change in medicine*. Cambridge, Mass. [etc.]: The MIT Press.

Blumenthal, D. (2003). Academic-industrial relationships in the life science. *The New England Journal of Medicine, 349*(25), 2452-24459.

Blumenthal, D., Campbell, E., Causino, N., & Louis, K. S. (1996). Participation of life-science faculty in research relationships with industry. *The New England Journal of Medicine, 335*(23), 1734-1739.

Bocking, S. (1997). *Ecologists and environmental politic: a history of contemporary ecology*. New Haven [etc.]: Yale University Press.

Boekholt, P., Meijer, I., & Vullings, W. (2001). *Evaluation of the valorisation activities of the Netherlands genomics initiative (NGI)*. Amsterdam: Technopolis.

Boogerd, F. C. & Westerhoff, H. V. (2007). *Systems biology: philosophical foundations*. Amsterdam [etc.]: Elsevier.

Borst, P. (2004). Knot door bureaucratie. *NRC Handelsblad*, April 3, 2004, p. 46.

Bowker, G. C. (2006). *Memory Practices in the Sciences*. Cambridge, MA [etc.]: MIT Press.

Bowker, G. C., & Star, S. L. (1999). *Sorting things out: classification and its consequences* Cambridge, MA [etc.]: MIT Press.

Braam, R. & Verbree, M. (2008). *Workshop research groups and science collaboration*. Working Report Series 1: Research groups. The Hague: Rathenau Institute, Science Systems Assesment.

Brand, S. (1994). *How buildings learn: what happens after they're built*. New York, NY [etc.]: Penguin books.

Breit, T. (2006). *E-biolab: an e-science laboratory for data analysis of omics experiments*. Paper presented at the Genomics: ready for the next step, Rotterdam.

Brown, N., Faulkner, A., Kent, J., & Michael, M. (2006). Regulating Hybrids: 'Making a Mess' and 'Cleaning Up' in Tissue Engineering and Transpecies Transplantation. *Social theory & health*, 4 (1), 1-24.

Brown, N., Rappert, B., & Webster, A. (2000). *Contested futures: a sociology of prospective techno-science*. Aldershot: Ashgate.

Brown, N., & Webster, A. (2004). *New medical technologies and society: reordering life*. Cambridge [etc.]: Polity.

Bucchi, M. (2004). Science in society: an introduction to social studies of science (A. Belton, Trans.). London [etc.]: Routledge.

Bush, V. (1980). *Science, the endless frontier: a report to the President*. New York: Arno Press.

Butler, J. M. (2005). *Forensic DNA typing: biology, technology, and genetics of STR markers* (2nd ed.). Amsterdam [etc.]: Elsevier.

Calvert, J. (2006). What's Special about Basic Research? *Science, technology, & human values, 31*(2), 199.

Cantor, C. R. (1990). Orchestrating the Human Genome Project. *Science, 248*(4951), 49-51.

Capshew, J. H. & K. A. Rader (1992). Big Science: Price to Present. *Osiris, 7*(Science after '40), 2-25.

Castells, M. (1996). *The rise of the network society.* Oxford: Blackwell.

Chadarevian, S. de (2002). *Designs for life: molecular biology after World War II.* Cambridge [etc.]: Cambridge University Press.

Chang, H. (1999). *Surviving the SOC revolution: a guide to platform-based design.* Boston; Dordrecht [etc.]: Kluwer Academic Publishers.

Check, E., & Castellani, F. (2004). David versus Goliath. *Nature, 432*(7017), 546-548.

Chompalov, I., Genuth, J., & Shrum, W. (2002). The organization of scientific collaborations. *Research Policy, 31*(5), 749-767.

Chompalov, I., & Shrum, W. (1999). Institutional Collaboration in Science: A Typology of Technological Practice. *Science, technology, & human values, 24*(3), 338-372.

Chong, L. & Ray L. B. (2002). Systems Biology - Whole-istic Biology. *Science, 295*(5560), 1661.

Cicmil, S. & Hodgson, D. (2006). *Making projects critical.* Basingstoke: Palgrave MacMillan.

Clarke, A. E., & Fujimura, J. H. (1992). *The right tools for the job: at work in twentieth century life sciences.* Princeton: Princeton University Press.

Collins, F. S., & Green. E.D. (2003). A vision for the future of genomics research. *Nature, 422*(6934), 835 -848.

Cook-Deegan, R. M. (1994). *The gene wars: science, politics, and the human genome.* New York, N.Y., [etc.]: Norton.

Cortright, J., & Mayer, H. (2002). *Signs of Life. The Growth of Biotechnology Centers in the U.S.* Washington DC: The Brookings Institution Center on Urban and Metropolitan Policy.

Crane, D. (1972). *Invisible colleges: Diffusion of knowledge in scientific communities.* Chicago: University of Chicago Press.

Creager, A. N. H. (1998). Biotechnology and Blood: Edwin Cohn's Plasma Fractationation Project, 1940-1953. In A. Thackray (Ed.), *Private science: biotechnology and the rise of the molecular sciences* (pp. 39-62). Philadelphia, PA: University of Pennsylvania Press.

Creager, A. N. H., & Santesmases, M. J. (2006). Radiobiology in the Atomic Age: Changing Research Practices and Policies in Comparative Perspective. *Journal of the History of Biology, 39*(4), 637.

Crease, R. P. (1999). *Making physics: a biography of Brookhaven National Laboratory, 1946-1972* Chicago, Ill [etc.]: The University of Chicago Press.

Cutcliffe, S. H. & Mitcham, C. (Ed.). (2001). *Visions of STS: Counterpoints in Science, Technology, and Society Studies.* Albany NY: SUNY Press.

Darwin, C. (1859). *On the Origin of Species by Means of Natural Selection, or the Preservation of Favoured Races in the Struggle for Life.* London: John Murray

Davis, B., & colleagues (1990). The human genome and other initiatives. *Science, 249*(4967), 342-343.

Decker, C., & O'Dor, R. (2003). A Census of marine life: unknowable or just unknown? *Oceanologica Acta*(25), 179-186.

Dijk, J. A. G. M. van (1991). *De netwerkmaatschappij: sociale aspecten van nieuwe media.* Houten [etc.]: Bohn Stafleu Van Loghum.

Disco, C. & B. van der Meulen (Ed.). (1998). *Getting new technologies together: studies in making sociotechnical order.* Berlin [etc.]: Walter de Gruyter.

Doing, P. (2008). Give me a laboratory and I will raise a discipline: the past, present, and future of politics of laboratory studies in STS In E. J. Hackett, O. Amsterdamska, M. Lynch & J. Wajcman (Eds.), *The Handbook of Science and Technology Studies-third edition* (pp. 279-296). Cambridge, MA: MIT.

Doorman, M. (2004). *De Romantische Orde.* Amsterdam: Uitgeverij Bert Bakker.

Douglas, C. (2005) Managing HuGE Expectations: Rhetorical Strategies in Human Genome Epidemiology. *Science Studies,* 18(2), 26-45.

Dreger, A. (2000). Metaphors and Morality in the Human Genome Project. In P. R. Sloan (Ed.), *Controlling Our Destinies* (pp. 155-184). Notre Dame, Indiana: University of Notre Dame Press.

Driel, R. van (2004). *SysBioNL; a Dutch programme for food and pharma.* Amsterdam: unpublished report.

Driel, R. van & Westerhoff, H. (2003). Systeembiologie levert al fantastische resultaten op. *Bionieuws, 2003*(18).

Drucker, P. F. (1947). *Big business.* London & Toronto: W. Heinemann.

Drucker, P. F. (1969). *The age of discontinuity: guidelines to our changing society.* London [etc.]: Heinemann.

Dunbar, M. J. (1971). Anatomy of an Expedition. *Geographical Review, 61*(1), 161-163.

Dupré, J., & O'Malley, M. (2005). Fundemental Issues in Systems Biology. *Bioessays,* 27, 1270-1276.

EC (2002). *Life sciences and biotechnology; a strategy for Europe.* Retrieved March 20, 2004 from http://ec.europa.eu/biotechnology/pdf/com2002-27_en.pdf.

EDA (2001). *Strategic Planning in the Technology-Driven World: A Guidebook for Innovation-Led Development.* Retrieved April 13, 2003 from http://12.39.209.165/ImageCache/EDAPublic/documents/pdfdocs/1g3_5f21_5fstratplan_2dtech_2epdf/v1/1g3_5f21_5fstratplan_2dtech.pdf

Edge, D. (1995). Reinventing the Wheel. In S. Jasanoff, G. E. Markle, L. C. Petersen & T. J. Pinch (Eds.), *Handbook of Science and Technology Studies* (pp. 3-23). Thousand Oaks: Sage.

Editorial Washington Post (2005). No fish story. *Washington Post,* Saturday August 13, 2005, p. A20.

Eichinger, N. (2007). European funding targets big biology. Metagenomics and variomics benefit from round of grants. *Nature, 445,* 8-9 (published online 4 January 2007),.

Elias, P. (2004). *Animals to be added to human genome project.* Retrieved August 6, 2004 from http://www.iol.co.za/general/avant_newsview.php?click_id=31&art_id=qw1086673141155 B225&set_id=1

Elzinga, A. (2004). Metaphors, models and reification in science and technology policy discourse. *Science as culture,* 13(1), 105-122.

ERASysBio Partners (2007). *Systems Biology in the European Research Area.* Draft Strategy Paper: ERA-NET for Systems Biology.

Eriksson, L. (2008). Standardising the Unknown: practicable pluripotency as doable futures. *Science as Culture,* 17(1), 57-69.

ESF (2007). *Systems Biology: a Grand Challenge for Europe.* Strassbourg: EuropeanScience Foundation.

Etzkowitz, H., & Leydesdorff, L. (2000). The dynamics of innovation: from National Systems and "Mode 2" to a Triple Helix of university-industry-government relations. *Research Policy, 29*(2), 109-123.

Etzkowitz, H., & Webster, A. (1998). *Capitalizing knowledge: new intersections of industry and academia.* Albany, NY: State University of New York Press.

Etzkowitz, H. & Leydesdorff, L. (2003). Can "The Public" Be Considered as a Fourth Helix in University-Industry-Government Relations? Report of the Fourth Triple Helix Conference. *Science & Public Policy, 30* (1), 55-61.

EZ (2002). *Closing the gap-innovation lecture 2002.* The Hague: Ministry of Economic Affairs.

Fay, C. N. (1912). *Big business and government* New York: Moffat, Yard & Co.

Fernández-Armesto, F. (2006). *Pathfinders: a global history of exploration.* Oxford [etc.]: Oxford University Press.

Finlayson, A.C. (1994). *Fishing for truth: a sociological analysis of Northern cod stock assessment from 1977-1990.* Social and Economic Studies, no. 52. Canada: Memorial University of Newfoundland, Institute of Social and Economic Research.

Fitter, A. (2003). *Lord Sainsbury opens £25m University building.* Retrieved December 8, 2005 from http://www.york.ac.uk/admin/presspr/pressreleases/sainsburybiology.htm

Fleck, L. (1981). *Genesis and development of a scientific fact.* Edited by Thaddeus J. Trenn & Robert K. Merton. Translated by Fred Bradley and Thaddeus J. Trenn; preface by Thomas S. Kuhn. Chicago: Chicago University Press.

Folstar, P. (2002). Editorial. *News@genomics.nl, 1*(3), 3.

Fujimura, J. H. (1996). *Crafting science: a sociohistory of the quest for the genetics of cancer.* Cambridge, MA [etc.]: Harvard University Press.

Fujimura, J. H. (2005). Postgenomic futures: translations across the machine-nature border in systems biology. *New Genetics and Society, 24*(2), 195-225.

Funtowicz, S.O., & Ravetz, J.R. (1993). Science for the Post-Normal Age. *Futures, 25*(7), 735-755.

Funtowicz, S.O., & Ravetz, J. R. (1994). Uncertainty, Complexity and Post-Normal Science. *Environmental toxicology and chemistry, 13*(12), 1881-1886.

Galison, P. (1992). Introduction: the many faces of big science. In P. Galison & B. Hevly (Eds.), *Big science: the growth of large-scale research.* Stanford: Stanford University Press.

Galison, P. (1997). *Image and logic: a material culture of microphysics.* Chicago: University of Chicago Press.

Galison, P. (2003). The collective author. In M. Biagioli & P. Galison (Eds.) (2003). *Scientific authorship: credit and intellectual property in science.* (pp. 325-358). London: Routledge.

Galison, P., & Hevly, B. (1992). *Big science: the growth of large-scale research.* Stanford: Stanford University Press.

Galison, P., & Thompson, E. (1999). *The architecture of science.* Cambridge: MIT Press.

Galison, P., & Jones, C. A. (1999). Factory, Laboratory, Studio: Dispersing Sites of Production. In P. Galison & E. Thompson (Eds.), *The architecture of science.* Cambridge: MIT Press.

García-Sancho, M. (2008). Sequencing as a way of work: a history of its emergence and *mechanization – from proteins to DNA, 1945-2000.* PhD thesis: Centre for the History of Science, Technology and Medicine, Imperial College, London.

Gaudillière, J.-P., & Rheinberger, H.-J. (Eds.). (2004). *From molecular genetics to genomics: the mapping cultures of twentieth century genetics.* London: Routledge.

Gibbons, M., Limoges, C., & Nowotny, H. (1994). *The new production of knowledge: the dynamics of science and research in contemporary societies.* London: Sage.

Gieryn, T. (1983). Boundary-Work and the Demarcation of Science from Non-Science: Strains and Interests in the Professional Ideologies of Scientists. *American Sociological Review, 48*, 781-795.

Gieryn, T. (1995). Boundaries of Science. In S. Jasanoff, G. E. Markle, L. C. Petersen & T. J. Pinch (Eds.), *Handbook of Science and Technology Studies* (pp. 293-443). Thousand Oaks: Sage.

Gieryn, T. F. (1999). *Cultural boundaries of science: credibility on the line.* London: University of Chicago Press.

Glasner, P. (1996). From community to 'collaboratory'? The Human Genome Mapping Project and the changing culture of science. *Science and public policy, 23*(2), 109-116.

Glasner, P. (2002). Beyond the genome: reconstituting the new genetics. *New Genetics and Society, 21*(3), 267-277.

Goffman, E. (1973). *The presentation of self in everyday life*. Woodstock, NY: Overlook Press.

Gombrich, E. (1968/72). Style. In D. L. Sills (Ed.), *International Encyclopedia of the Social Sciences* (Vol. XV, pp. 352-361). New York: Macmillan & The Free Press/London: Collier-Macmillan

Goodman, N. (1978). *Ways of worldmaking*. Indianapolis: Hackett.

Graham, L. R.. (1992). Big Science in the Last Years of the Big Soviet Union. *Osiris, 7*(Science after '40), 49-71.

Grassle, J. F. (1997). *Report to the Alfred P. Sloan Foundation: Workshop to consider the scientific and technical aspects of a census of marine benthic species*. Retrieved 28 September 2006, from http://marine.rutgers.edu/OBIS/origin/MtgRprt.htm

Grassle, J. F., & Stocks, K. I. (1999). A Global Ocean Biogeographic Information System (OBIS) for the Census of Marine Life. *Oceanography 12*(3), 12-14.

Griesemer, J. R., & Gerson, E. M. (1993). Collaboration in the Museum of Vertebrate Zoology. *Journal of the History of Biology, 26*(2), 185-203.

Grimm, D. (2004). 2004 Visualiszation Challenge: Illustration. *Science, 305*(5692), 1905.

Grimm, D. (2004). Ancestral Mammal's Genome Reconstructed. *ScienceNow, 1202*(1). Retrieved August 6, 2004 from http://sciencenow.sciencemag.org/cgi/content/full/2004/1202/1.

Groenewegen, P., & Wouters, P. (2004). Genomics, ICT and the formation of R&D networks. *New Genetics and Society, 23*(2), 167-185.

Guston, D. H. (2000). *Between politics and science: assuring the integrity and productivity of research*. Cambridge: Cambridge University Press.

Guston, D. H. (2001). Special issue: Boundary organizations in environmental policy and science *Science, technology, & human values, 26*(4), 399-531.

Hackett, E.J., (2005). Introduction to the Special Guest-Edited Issue on Scientific Collaboration. *Social studies of science, 35*(5), 667-672.

Hackett, E. J., Amsterdamska, O., Lynch, M. & Wajcman, J. (Eds.). (2008). *The handbook of science and technology studies* (3rd ed.). Cambridge, MA [etc.]: MIT Press.

Hall, C. T. (2003). Formula for scientific innovation: Omit walls. Design of Stanford's Clark Center fosters interdisciplinary research. *San Francisco Chronicle*, October 20, 2003. Retrieved November 10, 2006 from http://www.sfgate.com/cgi-bin/article.cgi?f=/chronicle/archive/-2003/10/20/MNG9V2F3DL1.DTL

Health Protection Agency (2006). Influenza Pandemics of the 20th century. Retrieved October 29, 2007 from http://www.hpa.org.uk/infections/topics_az/influenca/pandemic/history.-htm

Hedström, P. (2006). Experimental macro sociology: predicting the next best seller. *Science*, 311(5762), 786 - 787.

Heijden, M. van der. (2008). De deeltjestsunami. Wereldwijd computernetwerk moet speld in CERN-hooiberg vinden. *NRC Handelsblad*, March 15&16, 2008, p. 37.

Held, D. (1999). *Global transformation: politics, economics and culture*. Cambridge: Polity Press in assoc. with Blackwell.

Hendrick, B. J. (1919). *The age of big business: a chronicle of the captains of industry* New Haven, CT [etc.] Yale University Press [etc.].

Hessels, L. & Lent, H. van (2008). Re-thinking new knowledge production: A literature review and a research agenda. *Research Policy, 37*, 740-760

Hevly, B. (1992). Afterword: reflections on big science and big history. In P. Galison & B. Hevly (Eds.), *Big science; the growth of large-scale research* Stanford: Stanford University Press.

Hilgartner, S. (1995). The Human Genome Project. In S. Jasanoff (Ed.), *Handbook of science and technology studies* (pp. 302-315). Thousand Oaks, CA [etc.]: Sage.

Hilgartner, S. (2000). *Science on stage: expert advice as public drama*. Stanford, CA: Stanford University Press.

Hine, C. (2006). Databases as Scientific Instruments and Their Role in the Ordering of Scientific Work. *Social studies of science, 36*(2), 269-298 (230).

Hoddeson, L., & Baym, G. (1993). *Critical assembly: a technical history of Los Alamos during the Oppenheimer years, 1943-1945*. Cambridge [etc.]: Cambridge University Press.

Hodgson, D. E. (2004). Project Work: The Legacy of Bureaucratic Control in the Post-Bureaucratic Organization *Organization: the interdisciplinary journal of organization, theory and society, 11*(1), 81-100

Holmes, B. (2004). 21st century ark: taking stock of nature's riches. *New Scientist*, 26 June 2004 (2453), 31-35.

Hood, L. (1990). No: and anyway, the HGP isn't Big Science. *The scientist, 4*(22), 13.

Hooke, R. (1987). *Micrographia: or some physiological descriptions of minute bodies made by magnifying glasses, with observations and inquiries thereupon*. Lincolnwood, IL: Science Heritage.

Hopkin, M. (2004). Decoders target 18 new genomes. *Nature*, published online August 4, 2004. Retrieved August 31, 2008 from http://www.nature.com/news/2004/040804/full/-news040802-12.html

Hopkin, M. (2005). Genome project aims to make Manhattan. 'Whole-environment sequencing' will reveal bugs in urban air. *Nature*, published online March 9, 2005. Retrieved August 31, 2008 from http://www.nature.com/news/2005/050307/full/news050307-12.html

Hubbard, J. (2003). *James H. Clark: Collaborative Environment key to innovation* [video]. Made for the Stanford Report, October 27. Retrieved November 20, 2006 from http://news-service.stanford.edu/news/2003/october29/jamesclark-video-1029.html.

Hull, D. L., & Ruse, M. (Eds.). (1998). *The Philosophy of Biology*. New York/Oxford: Oxford University Press.

International Polar Year. (2005). *History of IPY*. Retrieved February 14, 2007 from http://classic.ipy.org/development/history.htm

Irvine, J., & Martin, B. R. (1984). *Foresight in Science*. London: Pinter.

Jasanoff, S., Markle, G.E., Petersen, L.C. & Pinch, T.J. (Eds.) (1995). *Handbook of science and technology studies* (Revised paperback ed.). Thousand Oaks, CA [etc.]: Sage.

Jasanoff, S. (2005). *Designs on nature: science and democracy in Europe and the United States*. Princeton, NJ [etc.]: Princeton University Press.

Johnston, S., Skamene, E., Rima, B., & Melero, J. (2007). *External Research Review VIRGO Consortium 2006*. Retrieved October 29, 2007 from http://www.qanu.nl/comasy/uploadedfiles/-Virgo.pdf

Jungk, R. (1968). *Big machine*. New York: Scribner.

Kahmsi, R. (2004). Ancient mammal genes reconstructed. *Nature*, published online December 1, 2004. Retrieved August 31, 2008 from http://www.nature.com/news/2004/041129/full/-news041129-5.html

Katz, J. S., & Martin, B. R. (1997). What is research collaboration? *Research Policy, 26*, 1-18.

Kay, L. E. (2000). *Who wrote the book of life?: a history of the genetic code*. Stanford: Stanford University Press.

Keating, P & Cambrosio, A. (2003). *Biomedical platforms: realigning the normal and the pathological in late-twentieth-century medicine*. Cambridge, Mass. [etc.]: MIT Press.

Keller, E. Fox (1995). *Refiguring life: metaphors of twentieth-century biology*. New York: Columbia University Press.

Keller, E. Fox (2005). The century beyond the gene. *Journal of biosciences, 30*(1), 3-10 (18).

Kevles, D. J. (1992). Out of Eugenics: The historical politics of the human genome. In D. J. Kevles & L. E. Hood (Eds.), *The code of codes: scientific and social issues in the human genome project* (pp. 3-36). Cambridge MA [etc.]: Harvard University Press.

Kevles, D. J., & Hood, L. E. (1992). *The Code of codes: scientific and social issues in the human genome project.* Cambridge, MA. [etc.]: Harvard University Press.

Kleinman, D. L. (2003). *Impure culture: university biology and the world of commerce.* Madison, WI [etc.]: University of Wisconsin Press.

Kleinman, D. L., & Vallas, S. (2006). Contradiction in convergence: universities and industry in the biotechnology field. In S. Frickel & K. Moore (Eds.), *The new political sociology of science: institutions, networks, and power.* Madison, WI: The University of Wisconsin Press.

Knapen, B. (2007). De maakbare samenleving ligt in Azië. *NRC Handelsblad,* 20 januari 2007.

Knorr-Cetina, K. D. (1999a). *Epistemic cultures: how the sciences make knowledge.* Cambridge, Ma etc.: Harvard University Press.

Kohler, R. E. (1994). *Lords of the fly: Drosophila genetics and the experimental life.* Chicago, IL [etc.]: The University of Chicago Press.

Kolata, G. (1999). *Flu: The Story of the Great Influenza Pandemic of 1918 and the Search for the Virus That Caused It.* New York: Farrar, Straus and Giroux

Kumar, K. (1995). *From post-industrial to post-modern society: new theories of the contemporary world.* Oxford: Blackwell.

Kunzig, R. (2000). *Mapping the deep: the extraordinary story of ocean science.* New York [etc.]: Norton.

Kwa, C. (1987). Representations of Nature Mediating between Ecology and Science Policy: The Case of the International Biological Programme. *Social studies of science, 17*(3), 413-442.

Kwa, C. (2005). *De ontdekking van het weten. Een andere geschiedenis van de wetenschap.* Amsterdam: Boom.

Lamarck, J. B. (1809). *Zoological Philosophy: An Exposition with Regard to the Natural History of Animals* (H. Elliot, Trans.). Chicago: University of Chicago Press.

Lambright, H. W. (1998). Downsizing Big Science: Strategic Choices. *Public administration review, 58*(3), 259-268.

Lander, E. (2003a). *Beyond the Human Genome.* Paper presented at the scientific symposium From Double Helix to Human Sequence - and Beyond, Natcher Auditorium, NIH, April, 14, 2003. Retrieved August 31, 2008 from http://www.genome.gov/10506378

Lander, E. (2003b). *The Human Genome Project.* Paper presented at the public symposium From Double Helix to Human Sequence - and Beyond, Natural History Museum, Washington DC, April 15, 2003. Retrieved August 31, 2008 from http://www.genome.gov/10506379.

Lander, E. (2004). *Beyond the human genome.* Paper presented at the Genomics Momentum: Genomics for our world. Rotterdam, August 30 & September 1, 2004.

Latour, B. (1987). *Science in action: how to follow scientists and engineers through society.* Cambridge, Mass: Harvard University Press.

Latour, B. (1993). *We have never been modern.* Cambridge, MA: Harvard University Press.

Latour, B., & Woolgar, S. (1986). *Laboratory Life: the construction of scientific facts.* Princeton: Princeton University Press.

Law, J. (1994). *Organizing modernity* Oxford [etc.]: Blackwell.

Lederberg, J. & Uncapher, K. (Ed.). (1989). *Towards a national collaboratory.* Unpublished report of a National Science Foundation invitational workshop. New York: Rockefeller University.

Lenoir, T., & Hays, M. (2000). The Manhattan Project for Biomedicine. In P. R. Sloan (Ed.), *Controlling Our Destinies* (pp. 19-46). South Bend, Indiana: University of Notre Dame Press.

Lente, H. van (1993). *Promising technology: the dynamics of expectations in technological developments.* Enschede: Universiteit Twente.

Leslie, M. (2002). Confluence of Ocean Info. *Science,* 298, p. 1685.

Leydesdorff, L., & Etzkowitz, H. (1998). Triple Helix of innovation: Introduction. *Science and public policy, 25*(6), 358-364.

Lock, D. (2003). *Project management* (8th revised edition). Aldershot: Gower Publishing.

Lynch, M. (1991). Laboratory Space and the technological Complex: An Investigation of Topical Contextures. *Science in context, 4*(1), 51-78.

Mack, G. S. (2004). Can complexity be commercialized? *Nature Biotechnology, 22,* 1223–1229.

Magner, L. N. (1994). *A history of the life sciences* (2nd ed.). New York [etc.]: Dekker.

Maienschein, J. (1991). Cytology in 1924: Expansion and Collaboration. In K. R. Benson & J. Maienschein (Eds.), *The Expansion of American Biology* (pp. 23-51). New Brunswick [etc.]: Rutgers University Press.

Maienschein, J. (1993). Why Collaborate? *Journal of the History of Biology, 26*(2), 167-183.

Malakoff, D. (2000). Marine Census: Grants kick off ambitious count of all ocean life. *Science, 288*(5471), 1575-1576.

MBC (2002). *MassBiotech 2010: Achieving Global Leadership in the Life-Sciences Economy.* Retrieved April 10, 2003 from http://www.massbiotech2010.org/MassBioTech2010Report.pdf

McIntyre, A. D. (2005). Big science in the seas. *Marine Pollution Bulletin, 50,* 791-792.

Menard, J. W. (1969). *An Oceanic quest: the International Decade of Ocean Exploration International Decade of Ocean Exploration.* Washington DC: National Academy of Sciences.

Merton, R. K. (1965/1993). *On the shoulders of giants: a Shandean postscript.* Chicago [etc.]: The University of Chicago Press/Original publication: New York: Free Press.

Midler, C. (1995). Projectification of the firm: the renault case. *Scandinavian Journal of Management,* 11(4), 363-357.

Misa, T. J. (2004). *Leonardo to the Internet: technology and culture from the renaissance to the present.* Baltimore: Johns Hopkins University Press.

Müller-Rockstroh, B. (2007). *Ultrasound Travels. The Politics of a Medical Technology in Ghana and Tanzania.* Maastricht: Maastricht Univerty Press.

Nature Editorial (2001). Post-genomic cultures. *Nature, 409* (6822), 545.

Nauta, L. W. (1984). Exemplarische bronnen van het westers autonomie-begrip. *Kennis en methode,* 4(3), 190-208.

Neushul, P. (1993). Science, Government, and the Mass Production of Penicillin. *Journal of the history of medicine and allied sciences, 48*(4), 371-395.

NGA (2002). *A Governor's Guide to Cluster-Based Economic Development.* Retrieved April 10, 2003 from http://www.nga.org/cda/files/AM02CLUSTER.pdf

NGI (2001). *The Netherlands Genomics Strategy. Strategic plan 2002 – 2006.* The Hague: NGI.

NGI (2003). Dutch government awards Euro 86 million to NGI initiatives. *News@genomics.nl,* 2(4), 11.

NGI (2005a). *Genomics 2008-2010: bouwen en benutten; De Nederlandse genomics infrastructuur 2008-1012.* Den Haag: NGI.

NGI (2005b). *Netherlands Genomics Initiative, Annual Report 2004.* The Hague: NGI.

NGI (2005c). *Ondersteunende brieven.* Den Haag: NGI.

NGI (2005d). *Resultaten Nationale Genomics Strategie 2002-2007.* Den Haag: NGI.

NGI (2005e). VIRGO provides tools for anti-viral strategies. *News @ genomics.nl, 4*(1), 4-5.

NGI (2006). *Strategic Plan genomics 2008-2012.* The Hague: NGI.

NGI (2007a). *Businessplan NGI 2008-1012; Munt uit genomics.* Den Haag: NGI.

NGI (2007b). Dutch Cabinet awards NGI Euro 271 million. *Newsflash*, published online September 18, 2007. Retrieved September 1, 2008 from http://www.genomics.nl/News%-20archive/18%20September%202007.aspx.

Ngubane, B. (2001). *A National Biotechnology Strategy for South Africa*. Retrieved September 1, 2008 from http://www.dst.gov.za/publications-policies/strategies-reports/reports/dst_biotechnology_strategy.PDF.

Nielsen, W. (1972). *The big foundations*. New York: Columbia Press.

NIH (2003). *Human Genome; From blueprint to you*. Bethesda: NIH.

Nowotny, H., Scott, P., & Gibbons, M. (2001). *Re-thinking science: knowledge and the public in an age of uncertainty*. Cambridge etc.: Polity.

NRC Committee on Solar-Terrestrial Research (1994). *A Space Physics Paradox: Why Has Increased Funding Been Accompanied by Decreased Effectiveness in the Conduct of Space Physics Research?* Washington DC: The National Academies Press.

O'Malley, M. A., & Dupré, J. (2005). Fundamental issues in systems biology. *BioEssays, 27*, 1270-1276.

Oreskes, N. (2003). A Context of Motivation: US Navy Oceanographic Research and the Discovery of Sea-Floor Hydrothermal Vents. *Social studies of science, 33*(5), 697-742.

Owen-Smith, J. (2003). From separate systems to a hybrid order: accumulative advantage across public and private science at Research One universities. *Research Policy, 32*(6), 1081-1104

Owen-Smith, J., Massimo, R., Fabio, P., & Walter, W. P. (2002). A Comparison of U.S. and European University-industry Relations in the Life Sciences. *Management science, 48*(1), 24-43.

Packer, K. & Webster, A. (1996). Patenting culture in science: reinventing the scientific wheel of credibility. *Science, technology and human values*, 21 (4), 427-453.

Parker, J. (2006). *Organizational collaborations and scientific integration: the case of ecology and the social sciences*. Arizona State University, PhD Thesis.

Penders, B. (2008). *From seeking health to finding healths. The politics of large-scale cooperation in nutrition science*. Maastricht: Maastricht University Press.

Penders, B., Horstman, K., & Vos, R. (2008). Walking the line between biology and computation: the 'moist' zone. BioScience 58 (8), 747-755.

Pennisi, E. (2004). Ice Ages Put the Vice on Bison. *ScienceNow, 1129*(2).

Perkel, J. M. (2006). A Lab Startup. *The Scientist 20* (10), 75.

Pestre, D. J. K. (1992). Some thoughts on the early history of CERN. In P. Galison & B. Hevly (Eds.), *Big science; the growth of large-scale research* (pp. 78-99). Stanford CA: Stanford University Press.

Pickering, A. (1989). Pragmatism in particle physics: scientific and military interests in the post-war United States. In F. A. L. T. James (Ed.), *The Development of the Laboratory: Essays on the Place of Experiment in Industrial Civilization* (pp. 174-183). London: Macmillan.

Pickstone, J. V. (1993). Ways of knowing: towards a historical sociology of science, technology and medicine. *The British journal for the history of science, 26*(91), 433-458.

Pickstone, J. V. (2000). *Ways of knowing: a new history of science, technology and medicine*. Manchester: Manchester University Press.

Pickstone, J.V. (2007). Working knowledges before and after circa 1800. Practices and disciplines in the history of science, technology, and medicine. *Isis*, 98, 489-516.

Pirie, N. W. (1967). The part the international biological programme will play in increasing world food supplies. *The Proceedings of the Nutrition Society*, 26(1), 125-128.

Plasterk, R. (2003). Het gevaar dat systeembiologie heet. *Bionieuws, 2003*(16), 10.

PMP Public Affairs Consulting (1999). Biotechnology in Hawaii: A Blueprint for Growth. Retrieved April 20, 2003 from http://www.hawaii.gov/dbedt/info/energy/publications/-biotech99.pdf

Porter Liebeskind, J., Oliver, A. L., Zucker, L. G., & Brewer, M. B. (1995). *Social networks, learning, and flexibility: sourcing scientific knowledge in new biotechnology firms*. NBER: Working Paper 5320.

Powell, W. W., White, D.R., Koput, K.W. & Owen-Smith, J. (2005). Network Dynamics and Field Evolution: The Growth of Inter-organizational Collaboration in the Life Sciences. *American journal of sociology, 110*(4), 1132-1205.

Powell, W. W., & Owen-Smith, J. (2002). The new world of knowledge production in the life sciences. In S. G. Brint & C. Ker (Eds.), *The future of the city of intellect; the changing american university.* Stanford CA: Stanford University Press.

Power, M. (1997). *The Audit Society: Rituals of Verification.* Oxford: Oxford University Press.

Price, D. J. de Solla (1963). *Little science, big science.* New York [etc.]: Columbia University Press.

Pusey, M. J. (1945). *Big government: can we control it?* New York & London: Harper & Bros.

Rajan, K. Sunder (2003). Genomic Capital: Public Cultures and Market Logics of Corporate Biotechnology. *Science as culture, 12*(1), 87-122.

Rasmussen, N. (2002). Of 'small men', big science and bigger business: the second world war and biomedical research in the United States. *Minerva, 40*, 115-146.

Rechsteiner, M. (1990). The Human Genome Project: two points of view. *FASEB Journal, 4*(11), 2941-2942.

Remington, J. A. (1988). Beyond Big Science in America: The Binding of Inquiry. *Social studies of science, 18*(1), 45-72.

Rheinberger, H.-J. (1997). *Toward a history of epistemic things: synthesizing proteins in the test tube.* Stanford, Calif.: Stanford University Press.

Rickards, L. (2006). *Marine Data - A Big Issue.* Lecture during ICES annual scientific meeting, Maastricht, September 22, 2006.

Rifkin, J. (1999). *The biotech century: how genetic commerce will change the world.* London: Phoenix.

Rijck, K. de (2005). Griepdreiging of bangmakerij. *De Standaard,* September 2, 2005. Retrieved December 1, 2005 from http://www.standaard.be /artikel/printartikel.aspx?artikelId-=GEFHGDUS.

Rip, A. (1998). Fashions in science policy. In C. Disco & B. van den Meulen (Eds.), *Getting new technologies together: studies in making sociotechnical order.* Berlin: Walter de Gruyter.

Rip, A. (2001). Science for the 21st Century. In P. Tindemans, A. Verrijn-Stuart & R. Visser (Eds.), *The Future of the Sciences and Humanities; Four analytical essays and a critical debate on the future of scholastic endeavour.* Amsterdam: Amsterdam University Press.

Roberts, L. (2001). Controversial from the start. *Science, 291*(5507), 1182-1188.

Rogers, D. (1971). *The management of big cities; interest groups, and social change strategies.* Beverly Hills, CA: Sage publications.

Rose, N. (2001). The Politics of Life Itself. *Theory, culture & society, 18*(6), 1-30.

Sahlin-Andersson K. & Söderholm, A. (eds.). (2002). *Beyond project management; new perspectives on the temporary-permanent dilemma.* Copenhagen: Liber.

Salganik, M. J. & Dodds, P. S. et al. (2006). Experimental Study of Inequality and Unpredictability in an Artificial Cultural Market. *Science,* 311(5762), 854 - 856.

SARA (2006). *Virtual Reality.* Information leaflet. Amsterdam: SARA.

Schloegel, J.J. & Rader, K. A. (2005). *Ecology, Environment, and 'Big Science': An Annotated Bibliography of Sources on Environmental Research at Argonne National Laboratory, 1955-1985.* Argonne, Il:Argonne National Laboratory.

Schneiderman, B. (2008). Science 2.0. *Science,* 319(March 7), 1349-1350.

Schueler, J., Fickers, A., & Hommels, A. (Eds.). (2008). *Bargaining Norms Arguing Standards; Negotiating Technical Standards.* The Hague: Stichting Toekomstbeeld der Techniek.

Schumacher, E. F. (1973). *Small is beautiful: economics as if people mattered.* New York: Harper & Row.

Schwartz, M. (2003). Creative 'cauldrons' of research brew at new Clark Center. *Stanford Report,* October 22, 2003. Retrieved February 14, 2007 from http://news-service.stanford.edu/-news/2003/october22/xclark-1022.html.

Seidel, R. (1992). The origins of the Lawrence Berkeley Laboratory. In P. Galison & B. Hevly (Eds.), *Big science; the growth of large-scale research*. Stanford: Stanford University Press.

Shinn, T. (2002). The Triple Helix and New Production of Knowledge: Prepackaged Thinking on Science and Technology. *Social Studies of Science, 32* (4), 599-614.

Shreeve, J. (2004a). Craig Venter's Epic Voyage of Discovery. The epic quest to collect the DNA of everything on the planet - and redifine life as we know it. *Wired, 12*(8), 104-113/146-151.

Shreeve, J. (2004b). *The genome war: how Craig Venter tried to capture the code of life and save the world.* New York: Knopf.

Shrum, W., Genuth, J., & Chompalov, I. (2007). *Structures of scientific collaboration.* Cambridge, MA: MIT Press.

Sibal, K. (2005). *National biotechnology development strategy.* Dehli: Department of Biotechnology, Ministry of Science & Technology, Government of India.

Siebes, A. P. J. M. (2001). *Databases + data mining = bioinformatica.* Utrecht: Universiteit Utrecht.

Sismondo, S. (2004). *An introduction to science and technology studies.* London: Blackwell.

Sklair, L. (1973). *Organized knowledge: a sociological view of science and technology.* St. Albans: Hart-Davis MacGibbon.

Sloan, P. R. (2000). *Controlling our destinies: historical, philosophical, ethical, and theological perspectives on the Human Genome Project.* Notre Dame, IN: University of Notre Dame Press.

Sloterdijk, P. (2005). *Inspiration.* Paper presented at the lecture series Breath-taking. Air, art, architecture. Jan van Eyck Academie, Maastricht, May 31, 2005.

Smith, R. W. (1992). The biggest kind of big science: astronomers and the space telescope. In P. Galison & B. Hevly (Eds.), *Big science; the growth of large-scale research* (pp. 184-211). Stanford, California: Stanford University Press.

Snelgrove, P., & Grassle, F. (1995). What of the deep sea's future diversity? *Oceanus: the international magazine of marine science and policy, 38*(2), 29.

Snoep, J. L., Driel, R. van, & Westerhoff, H. V. (2003). *SiC! A Dutch initiative for an international Silicon Cells program.* Retrieved April 30, 2007 from http://www.systemsbiology.net/sbnl/-TSB.htm.

Star, S. Leigh, & Griesemer, J. R. (1989). Institutional Ecology, 'Translations' and Boundary Objects: Amateurs and Professionals in Berkeley's Museum of Vertebrate Zoology. *Social studies of science, 19*(3), 387-421.

Stehr, N. (Ed.). (2004). *The governance of knowledge* New Brunswick, N.J. [etc.] Transaction.

Stemerding, D., & Hilgartner, S. (1998). Means of coordination in making biological science: on the mapping of plants, animals and genes. In B. van der Meulen & C. Disco (Eds.), *Getting new technologies together: studies in making sociotechnical order* (pp. 39-69.). Berlin: Walter de Gruyter.

Sterckx, S. (1997). *Biotechnology, patents and morality.* Aldershot: Ashgate.

Stevens, R. C. (2004). Commentary - Long live structural biology. *Nature structural & molecular biology,* 11(4), 293-295.

Stoeckle, M., Bucklin, A., Knowlton, N., & Hebert, P. (2003). *Census of Marine Life DNA Barcoding Protocol.* Retrieved August 23, 2007 from http://www.comlsecretariat.org/Dev2Go.web?id=255158

Strasser, B. J. (2003a). The transformation of biological sciences in post-war Europe. EMBO and the early days of European molecular biology research. *EMBO reports, 4*(6), 540-543.

Strasser, B., J. (2003b). Who cares about the double helix? *Nature, 422*(6934), 803.

Sulston, J., & Ferry, G. (2002). *The common thread: a story of science, politics, ethics and the human genome.* London [etc.]: Bantam Press.

Taskforce on Systems Biology (2007). *ESF Taskforce on Systems Biology. Strategic Guidance and Recommendations.* Strassbourg: European Science Foundation.

Thacker, E. (2005). *The global genome: biotechnology, politics, and culture*. Cambridge, Mass. [etc.]: MIT Press.

Thackray, A. (1998). *Private science: biotechnology and the rise of the molecular sciences*. Philadelphia, PA: University of Pennsylvania Press.

Theberge, A. (2006). History of Ocean Exploration. *Ocean Explorer*. Washington DC: National Oceanic and Atmospheric Administration. Retrieved August 23, 2007 from http://ocean-explorer.noaa.gov/history/history.html

Tornatzky, L. G. (2000). *Building State Economies by Promoting University-Industry Technology Transfer*. Retrieved April 20, 2003 from http://www.nga.org/cda/files/UNIVERSITY.PDF

Traweek, S. (1988). *Beamtimes and Lifetimes: The World of High Energy Physicists*. Cambridge MA [etc.]: Harvard University Press.

Vanden Berghe, E., Costello, M. J., Guinotte, J., Fautin, D., & Halpin, P. (2006). *Marine Biology meets Information Technology – new era in marine biology begins*. Press release. Retrieved August 23, 2007 from http://www.marbef.org/modules.php?name=News&file=article&sid=119

Vaughan, D. (1999). The Role of the Organization in the Production of Techno-Scientific Knowledge. *Social studies of science, 29*(6), 913-944.

Venter, J. C. (2005). What can the oceans tell us about new energy development? *BioVision 2005: The World Life Sciences Forum Proceedings*. Lyon: Caliscope.

Venter, J. C. (2007). *A Life Decoded. My Genome: My Life*. New York: Viking.

Venter, J. C., Remington, K., Heidelberg, J. F., Halpern, A. L., Rusch, D., Eisen, J. A., et al. (2004). Environmental Genome Shotgun Sequencing of the Sargasso Sea. *Science, 304*(5667), 66-73.

Vermeulen, N., & Kleinenberg, R. (2003). Life Sciences in de Verenigde Staten. *TWAnieuws, 41*(4), 8.

Verne, J. (1870). *Vingt mille lieues sous les mers*. Paris: Hetzel.

Vliet, F. v. (2004). *Genomics is teamwork* [dvd]. Rotterdam: Erasmus MC.

Vriend, G. (2000). *Bioinformatics by trial and error*. Nijmegen: Katholieke Universiteit Nijmegen.

Vries, G. de (1981). De besmettelijkheid van wetenschappelijk contact. Bespreking van L. Fleck, Entstehung und Entwicklung einer wissenschaftlichen Tatsache. *Kennis en Methode, 5*, 156-164.

Vullings, W. & Planque, K. (2002). Genoomonderzoek in de VS. Aandacht voor maatschappelijke aspecten. *Technieuws, 5*, 26-41.

Wagner, C. S. (2004). *International collaboration in science: a new dynamic for knowledge creation*. Amsterdam: University of Amsterdam.

Waldby, C. & Mitchell, R. (2006). *Tissue economies: blood, organs, and cell lines in late capitalism*. Durham, NC [etc.]: Duke University Press.

Watson, J. D. (1968). *The double helix: a personal account of the discovery of the structure of DNA*. New York: Atheneum.

Weber, M. (1985). *The Protestant Ethic and the Spirit of Capitalism* (T. Parsons, Trans.). New York: Scribner.

Webster, A. (2005). Social science and a post-genomic future: alternative readings of genomic agency. *New Genetics and Society, 24*(2), 227-239.

Webster, A. & Eriksson, L. (2008). Governance-by-standards in the field of stem cells: managing uncertainty in the world of basic innovation. *New Genetics and Society, 27*(2), 99-111.

Weinberg, A. M. (1961). Impact of Large-Scale Science on the United States: Big science is here to stay, but we have yet to make the hard financial and educational choices it imposes. *Science, 134*(3473), 61-164.

Weinberg, A. M. (1967). *Reflections on big science*. Oxford etc.: Pergamon Press.

Weinberg, A. M. (1999). The birth of Big Biology. *Nature, 401*(6743), 738.

Westerhoff, H., & Palson, B. O. (2004). The evolution of molecular biology into systems biology. *Nature biotechnology, 22*(10), 1249-1252.

Westfall, C. (2003). Rethinking Big Science: Modest, Mezzo, Grand Science and the Development of the Bevalac, 1971-1993. *Isis, 94*(1), 30-56.

Westwick, P. J. (2003). *The national labs: science in an American system, 1947-1974.* Cambridge MA [etc.]: Harvard University Press.

Whitley, R. (1984). *The intellectual and social organization of the sciences.* Oxford: Clarendon Press.

WHO (2005a). *Avian Influenza: assessing the pandemic threat.* Retrieved October 29, 2007 from http://www.who.int/csr/disease/influenza/H5N1-9reduit.pdf

WHO (2005b). *WHO Global Influenza Preparedness Plan.* Geneva: WHO.

Wilde, R. de (2001). *De kenniscultus: Over nieuwe vormen van vooruitgangsgeloof.* Maastricht: Maastricht University Press.

Wiley, H. S. (2006). Systems Biology - Beyond the Buzz. *The Scientist,* June 2006, 53-57.

Wolthuis, A. (2006). Kiezen voor leasen. *Laboratorium magazine, 42*(8), 17-18.

Worm, B., Sandow, M., Oschlies, A., Lotze, H. K., & Myers, R. A. (2005). Global patterns of predator diversity in the open oceans. *Science, 309,* 1365-1369.

WRR (2003). *Beslissen over biotechnologie.* Den Haag: Sdu Uitgevers.

Wulf, W. (1989). The national collaboratory. In J. Lederberg & K. Uncapher (Eds.), *Towards a national collaboratory.* Unpublished report of a National Science Foundation invitational workshop. New York: Rockefeller University

Wyatt, S. (1998). *Technology's Arrow, Developing Information Networks for Public Administration in Britain and the United States.* Maastricht: Maastricht University Press.

Yoxen, E. (1984). *The gene business.* New York: Harper & Row Publishers.

Zallen, D. (1992). The Rockefeller Foundation and Spectroscopy research: The programs at Chicago and Utrecht. *Journal of the History of Biology, 25*(1), 67-89.

Zucker, L., G., Darby, M. R., & Armstrong, J. S. (2001*). Commercializing Knowledge: University Science, Knowledge Capture, and Firm Performance in Biotechnology.* Working Paper, NBER working paper series.

Zucker, L. G. (1995). *Collaboration structure and information dilemmas in biotechnology: organizational boundaries as trust production.* Cambridge, MA: National Bureau of Economic Research.

Zuidam, J. (2005). *Ondersteunende brieven.* Den Haag: NGI.

Appendix A

Meetings

2003

AAAS Colloquium on Science and Technology Policy, Washington D.C., USA, April 10-11.

Seminar series of the Biotechnology and Public Policy Initiative organised by the Center for Strategic and International Studies (CSIS), Washington D.C., USA, spring 2003.

Biotechnology: past, present and future, MIT on the road, National Academy of Sciences, Washington D.C., USA, April 12.

Scientific symposium 'From Double Helix to Human Sequence and Beyond', National Institutes of Health, Washington D.C., USA, April 14-15.

Public symposium 'Bringing the Genome to You', Natural History Museum, Washington DC., USA, April 15.

Symposium 'A Transatlantic Dialogue on Genetics and Health: Research Frontiers and Ethical, Economic, Legal, and Social Issues', Royal Norwegian Embassy & CSIS, Washington D.C., USA, May 16.

BIO 2003, Annual convention of the Biotechnology Industry Organization, Washington D.C., USA, June 22-25.

2004

AAAS Policy Forum, Washington DC, Washington D.C., USA, April 22 & 23.

Symposium 'Biodiversity in a changing world', Royal Academy of Arts and Sciences (KNAW), Amsterdam, The Netherlands, June 3.

Genomics Momentum, annual conference of the Netherlands Genomics Initiative, Rotterdam, The Netherlands, August 30-September 1.

Visit to Saragene at SARA Computing and Networking Services organized by the Genomics Network for Young Scientists (GeNeYouS), Amsterdam Science Centre, The Netherlands, September 20.

Theme day 'Entrepreneurs in the knowledge sector – Critical success factors for RTD intensive SMEs and starter companies in ICT, Nano- and Biotechnology', organised by the TWA network at Science Park, Eindhoven, The Netherlands, October 14.

GeNeYouS lecture and general meeting, Utrecht University, The Netherlands, October 27.

BIOnale 'Creating Life Science value in Europe', Maastricht, The Netherlands, November 2-4.

2005

Innogen conference 'Evolution of the Life Science Industries Conference', ESRC Centre for Social and Economic Research on Innovation in Genomics, Edinburgh, UK, 23-25 February.

4th World Life Sciences Forum BioVision, Lyon, France, April 11-15.

Participation in 'Class of 2005' at the World Life Science Forum BioVision, Lyon, France, April 10-15.

2nd Geneyous Symposium 'Interaction in Genomics', Organon, Oss, the Netherlands, April 27.

Symposium 'Towards a Philosophy of Systems Biology' Free University, Amsterdam, The Netherlands, June 2-3.

Egenis conference 'Genomics in Context' ESRC Centre for Genomics in Society, University of Exeter, UK, September 28-30.

Reception Applied Genomics programme, a collaboration between the Biotechnology and Biological Sciences Research Council (BBSRC), the Medical Research Council (MRC) and the Department of Trade and Industry (DTI), London, UK, October 29.

2006

The third International Conference 'Genomics and Society: towards a socially robust science?' organised by the Centre for Society and Genomics (CSG) and the ESRC Centre for Economic and Social Aspects of Genomics (CESAGen), Amsterdam, 20-21 April.

Annual science conference of the International Council for the Exploration of the Sea (ICES), Maastricht, The Netherlands, September 19-23.

'Census of Marine Life: Community and species biodiversity in marine benthic habitats from the coastal zone to the deep sea', ICES annual science conference, Maastricht, The Netherlands, September 19.

10th Annual colloquium of the Postgraduate Forum on Genetics and Society (PFGS) 'Implications and Implementations: The Meaning and Use of our Research', York University, UK, August 30- September 1.

Genomics Momentum *'Genomics. Ready for the next step'*, Rotterdam, the Netherlands, November 9.

2007

11th annual PFGS colloquium 'Public Participation and Engagement in the Life Sciences', Grey College, University of Durham, UK, August 29-31.

APPENDIX B

Overview interviews

Andeweg, Dr. Arno. Coordinator VIRGO, Department of Virology, Erasmus Medical Centrum, Rotterdam, The Netherlands. Rotterdam: April 29, 2005.

Berg, Dr. Hans van den. Research coordinator pharma, Organon, Oss, The Netherlands. Oss: March 21, 2005.

Bie, dr. Rien de. Research manager, Bijvoet Centre, Utrecht University, The Netherlands. Utrecht: November 10, 2006.

Bos, Prof. Dr. Kors. Computation coordinator of the ATLAS experiment, CERN, Geneva, Suisse. Geneva: April 30, 2008.

Bowles, Prof. Dr. Dianna, Director of the Center for Novel Agricultural Products and Weston Chair of Biochemistry, Biology Department York University, UK. York: August 30, 2006.

Bunthof, Dr. Christine. Coordinator ERA-NET Plant Genomics, Netherlands Genomics Initiative, The Hague, The Netherlands. Cofounder GeNeYouS. Rotterdam: November 9, 2006.

Dons, Prof. Dr. Hans. Biopartner professorship on entrepreneurship in the life sciences at the University of Nijmegen and the University of Wageningen in the Netherlands. Director of Keygene B.V. Wageningen: August 11, 2004.

Dopson, Dr. Sue. Fellow in organizational behaviour, Said Business School, Oxford, UK. Oxford: February 24, 2006.

Driel, Prof. Dr. Roel van. Professorship in Biochemistry, Swammerdam Institute for Life Sciences, University of Amsterdam, The Netherlands. Amsterdam: March 30, 2005.

Eijl, Dr. Henriette van. Policy officer at Directorate F (Health) and Unit F4 (Fundamental genomics), Seconded National Expert, DG Research, European Commission, Brussels, Belgium. Brussels: April 19, 2005.

Geus, Dr. Bernard de. Project coordinator, Netherlands Genomics Initiative, The Hague, The Netherlands. The Hague: March 16, 2005

Heip, Prof. dr. Carlo. Director of the Centre for Estuarine and Marine Ecology of the Netherlands Institute of Ecology (NIOO-KNAW), Yerseke, The Netherlands. General director of the Royal Netherlands Institute of Sea Research (NIOZ), Texel, The Netherlands. Member of the Scientific Steering Committee of CoML and founder of the Euro-COML committee. Yerseke: December 11, 2006.

Horning, Menno. Head of the life sciences department, Ministry of Economic Affairs, The Hague, The Netherlands. The Hague: February 21, 2003.

Hubbard, Prof. Dr. Rod, Director of the Structural Biology Laboratory, York University. York: December 22, 2005.

Lelyveld, Dr. Philip van. Director Public Affairs, DSM Life Sciences, Delft, The Netherlands. Delft: February 12, 2003.

McCarthy, Prof. Dr. John. Director of the Manchester Interdisciplinary Biocentre (MIB), UK. Manchester: December 13, 2005.

Musselwhite, Bernadette. Specialist business development, Montgomery County Department of Economic Development, Maryland, USA. Washington DC: June 2, 2003.

Ogg, Ron. Director Biopartner Center Maastricht, The Netherlands. Maastricht: January 21, 2003.

Pierrot-Bults, Dr. Annelies. Senior researcher at the Zoölogical Museum of the University of Amsterdam, The Netherlands. Member of Mar-Eco, CMarZ and the bar-coding working group of CoML. Amsterdam: January 8, 2007.

Remacle, Dr. Jacques. Policy officer unit F4, DG research, European Commision, Brussels, Belgium. Brussels: October 6, 2006.

Rietveld, Dr. Luc. Policy officer, ZonMW and Netherlands Genomics Initiative, The Hague, The Netherlands. The Hague: December 12, 2006.

Sanders, Prof. Dr. Dale, Head of the Biology Department, York University. York: December 12, 2005.

Sibuet, Prof. Dr. Myriam. Director of the 'Département Environnement Profond' of the 'Institut français de recherche pour l'exploitation de la mer' (Ifremer), Brest, France. Member of the Scientific Steering Committee of CoML and programme leader of CoMarge. Maastricht, September 20, 2006.

Siebes, Prof. Dr. Arno. Professorship in informatics, specialised in large-scale distributed data systems at the Institute of Information and Computing Sciences, Utrech University, The Netherlands. Utrecht: July 29, 2004.

Sinclair, Prof. Dr. Michael. Director of the Bedford Institute of Oceanography (BIO), Darthmouth, Canada. Member Scientific Steering Comittee CoML and participates in the pilot project Gulf of Maine Census. Maastricht: September 20, 2006.

Spek, Dr. Wouter. Manager innovation and international affairs, Netherlands Genomics Initiative, The Hague, The Netherlands. The Hague: February 10, 2003.

Verschoor, Dr. Marga, Policy officer NWO-ALW, The Hague, The Netherlands. The Hague: January 5, 2007.

Vriend, Prof. Dr. Gert. Professorhip in Bioinformatics of Macromolecular Structures. Director of the Center for Molecular and Biomolecular Informatics at Nijmegen University. Nijmegen: August 13, 2004.

Weiss, Marc. Director Economic Development, Hanover County Department of Economic Development, Virginia, USA. Washington DC: June 24, 2003.

Westerhoff, Prof. Dr. Hans. Director of the Department of Molecular Cell Physiology at the Free University, Amsterdam, The Netherlands. Group leader of the systems biology research group of the Manchester Interdisciplinary Biocentre, UK. Amsterdam: June 27, 2005.

Woude, Dr. Marianne van der. Biology Department York University. York: December 15, 2005.

Zeegers, Prof. Dr. Maurice. Professorship in Genetic Epidemiology, The Department of Public Health and Epidemiology, The University of Birmingham, UK. Cofounder GeNeYouS. Leuven: August 9, 2006.

*All the interviews have been recorded. Quotes from these interviews have been authorised by the interviewees, except for the quotes in chapter 2 from Rien de Bie as he passed away shortly after the interview took place.

Curriculum Vitae

Niki Vermeulen (1978) has completed her PhD research on scientific collaboration in biology as a member of the Faculty of Arts and Social Sciences of Maastricht University, the Netherlands, first in the Department of Philosophy and later on in the Department of Technology and Society Studies. She has also taught several courses in this Faculty. As part of her PhD work she was enrolled in the graduate training programme of the Netherlands Graduate Research School of Science, Technology and Modern Culture (WTMC). In 2005 she was a Marie Curie research fellow at the Science and Technology Studies Unit at the University of York, UK.

Niki graduated in Arts and Culture at Maastricht University in 2003. She specialized in Science and Technology Studies and has written her Master thesis on the origin and application of social-cultural network theories. During her study, she served as a research-assistant in two projects: one about the knowledge society and one about the role of Non-Governmental Organisations in the discussion on agricultural biotechnology. In 2001 she participated in a project of the Netherlands Scientific Council for Government Policy (WRR), geared to analysing the influence of information and communication technology on knowledge policies and in 2003 she wrote a report on life sciences innovation in the United States, for the Scientific Council of the Royal Netherlands Embassy in Washington DC, USA. In the academic year 2008/2009 she has been working as a research policy advisor for the Executive Board of Maastricht University.

In addition to her research, Niki Vermeulen has been an active member of the scientific community. She served as a member of the University Council of Maastricht University, as a member of the council of the European Association for the Study of Science and Technology, as a member of the graduate school's programme committee, and she co-founded the Marie-Curie network BioStep. Currently, Niki is a visiting researcher and lecturer at the Department of Social Studies of Science of the University of Vienna, Austria, and she works as a consultant for the Technopolis group office in Vienna. In her future career she hopes to continue combining scholarly research and policy work.